PE POWER STUDY GUIDE

FOURTH EDITION

JOHN A. CAMARA, PE

PPI®

PPI2PASS.COM

A **KAPLAN** COMPANY

Report Errors for This Book

PPI is grateful to every reader who notifies us of a possible error. Your feedback allows us to improve the quality and accuracy of our products. Report errata at **ppi2pass.com**.

NFPA 70®, National Electrical Code®, and NEC® are registered trademarks of the National Fire Protection Association, Inc., Quincy, MA 02169. National Electrical Safety Code® and NESC® are registered trademarks of the Institute of Electrical and Electronics Engineers, Inc., New York, NY 10016.

PE POWER STUDY GUIDE
Fourth Edition

Current release of this edition: 2

Release History

date	edition number	revision number	update
Jan 2021	4	1	New edition. Copyright update.
Mar 2022	4	2	Minor corrections.

© 2022 Kaplan, Inc. All rights reserved.

Printed in the United States of America.

PPI
ppi2pass.com

ISBN: 978-1-59126-785-0

Table of Contents

Preface

In 2020, the National Council of Examiners for Engineering and Surveying (NCEES) changed the Professional Engineering (PE) electrical engineering licensing exams from a pencil-and-paper exam to a computer-based one. This in itself is not a big change. However, for the pencil-and-paper exam, you were permitted to bring your own reference material, including the *PE Power Reference Manual*, into the exam room. For the computer-based exam, you may not. Instead, you will have on-screen access to a searchable electronic copy of the *NCEES PE Electrical and Computer: Power Reference Handbook* (*NCEES Handbook*). This is the only reference you may consult during the exam—along with copies of select codes and standards, such as the *National Electrical Code*, which will also be provided.

This drastically changes how you must study for the exam. It is no longer enough to learn how to solve exam problems using the *PE Power Reference Manual* and other familiar reference books that you may annotate, highlight, and glue tabs to as you study. Now you must learn how to quickly find the equations and data you need in one specific source, an unmarked electronic copy of the *NCEES Handbook*.

That is the reason for this new book, the *PE Power Study Guide*. In addition to including detailed explanations of the equations and terms in the *NCEES Handbook*, this book bridges the gap between the *NCEES Handbook* and the *PE Power Reference Manual*.

Future editions of this book will be very much shaped by what you and others want to see in it. I am braced for the influx of comments and suggestions from those readers who (1) want more topics or (2) want more detail in existing topics. I appreciate suggestions for improvement, additional questions, and recommendations for expansion so that new editions will better meet the needs of future examinees.

I wish you all the best as you study for your exam. With the *PE Power Study Guide* as your road map and the *PE Power Reference Manual* and *NCEES Handbook* as your primary sources, I am confident you will have a study experience that is challenging but targeted to ensure success.

All the best. Enjoy the adventure of learning, creating, and building.

John A. Camara, PE

Acknowledgments

PE Power Study Guide required the development of a completely new kind of book. Accordingly, it was a team effort at PPI to design and publish this book. Their efforts were extensive, their patience outstanding, and their technical guidance absolutely instrumental. Many thanks to the staff at PPI, especially production manager Nicole Capra-McCaffrey, project manager Beth Christmas, product manager Nicole Evans, content specialist Meghan Finley, lead typesetter Richard Iriye, lead editor Scott Rutherford, and systems manager Sam Webster.

Content team: Bonnie Conner, Meghan Finley, Anna Howland, Amanda Werts

Editorial team: Bilal Baqai, Tyler Hayes, Indira Prabhu Kumar, Scott Marley, Scott Rutherford, Michael Wordelman

Editorial operations director: Grace Wong

Project manager: Beth Christmas

Product management team: Nicole Evans, Megan Synnestvedt

Production team: Kim Burton-Weisman, Nikki Capra-McCaffrey, Richard Iriye, Sean Woznicki, and Stan Info Solutions

Publishing systems manager: Sam Webster

Technical illustrations and cover design: Tom Bergstrom

Although I had help in preparing this new edition, the responsibility for any errors is my own. A current list of known errata for this book is maintained at the PPI website (**ppi2pass.com**), where you can also let me know of any errors you find. I greatly appreciate the time you take to help keep this book accurate and up to date.

Suggestions for improvement are appreciated. I hope this book serves you well. Expect future updates as NCEES refines the computer-based exam.

John A. Camara, PE

How to Use This Book

Congratulations on your decision to earn your PE! You have chosen a path that will open many doors to exciting and interesting opportunities.

The path to PE licensure is one of the most difficult and rewarding experiences in the life of an engineer. The adventure you've embarked on is an intense and rigorous one. It will be daunting; at times it will even feel impossible. However, once successful, you will stand out from your peers.

1. GOALS

The *PE Power Study Guide* (and the range of PPI publications that support it) is designed to prepare you for success on the exam. It is a bridge between the *NCEES Handbook* and the *PE Power Reference Manual*. Using the *PE Power Study Guide* and the in-depth information in the *PE Power Reference Manual*, you will gain a comprehensive understanding of the equations and concepts most likely to be encountered on the exam.

The primary focus of this book is the knowledge contained in the *NCEES Handbook*, which is the only reference you will have access to during the exam. However, past experience—and the NCEES's own exam specifications—show that there are a multitude of key concepts required for success on the exam which are not included in the *NCEES Handbook*—Kirchhoff's voltage law and ladder logic, for example. With this in mind, the *PE Power Study Guide* has been written with three main goals.

1. **To ensure that you study exactly what you need to study—no more and no less.** In addition to providing detailed explanations of *NCEES Handbook* equations and content, the *PE Power Study Guide* also covers key concepts which are not included in the *NCEES Handbook*.

2. **To guide you in using the *NCEES PE Electrical and Computer: Power Reference Handbook* (often shortened to *NCEES Handbook* or simply *Handbook*) with maximum efficiency.** Because this is the only reference guide available to you (other than the NEC and select other codes) during the exam, it is crucial you are able to reference it with ease and confidence. This book is your guide to such confidence.

3. **To show you where to look for more vital information.** In most cases, this is the *PE Power Reference Manual* book.

2. CHAPTER STRUCTURE

The chapters of this *PE Power Study Guide* follow the NCEES exam specifications. Each of the nine NCEES exam specifications is covered by one of the nine chapters in this book.

Each chapter covers all the knowledge areas in the corresponding NCEES exam specification. The organization of each chapter follows the order of knowledge areas given by the NCEES for the related exam specification.

The following diagram illustrates the knowledge areas for exam specification 2.A, Analysis (Chapter 4 in this book).

2. Circuits — exam specification (chapter level)
   A. Analysis
      1. Three-Phase Circuits
      2. Symmetrical Components
      3. Per Unit System
      4. Phasor Diagrams — knowledge areas (chapter subheadings)
      5. Single-Phase Circuits
      6. DC Circuits
      7. Single-Line Diagrams

Go to NCEES.org and download a PDF of the exam specifications so you can reference it while you study.

Each chapter is organized as follows. (Boldface indicates the actual section title.)

1. information about the exam specification

2. knowledge areas related to the exam specification

   (a) **Knowledge Area Overview** for each knowledge area

   (b) relevant equations and concepts for each knowledge area

3. **Definitions**

4. **Nomenclature**

3. KNOWLEDGE AREA OVERVIEW

In the **Knowledge Area Overview** section at the beginning of each knowledge area, there are three lists.

1. The key concepts you need to know to successfully answer the exam questions related to the knowledge area.

2. All the sections in EPRM where you can find information on this knowledge area. If only the chapter is listed, refer to all sections in that chapter.

3. All the sections in the *NCEES Handbook* where you can find the relevant equations and concepts for this knowledge area.

4. RELEVANT EQUATIONS AND CONCEPTS

The equations and relevant concepts you will need to know for each exam specification are presented in the related knowledge area sections. Whenever possible, the relevant *NCEES Handbook* section titles are cited for an equation or concept. The equation and section numbers from the *PE Power Reference Manual* are also cited, so you can easily find more in-depth information.

Blue content: Equations and text in blue are directly from the *NCEES Handbook*. For each knowledge area, *blue* equations are covered in the same order as they appear in the *NCEES Handbook* sections.

Red content: Equations and titles in red are essential knowledge for the exam. These are the equations from the *PE Power Reference Manual* you will need to memorize or derive for success on exam day.

For equations that are identical or nearly identical in presentation in both the *NCEES Handbook* and the *PE Power Reference Manual* (EPRM), you will see a single blue equation with a reference to its location in each book. Some *NCEES Handbook* equations, however, differ substantially in presentation from the same equation in EPRM. For example, the *NCEES Handbook* equation may use a different variable, different subscript, or may be a slightly rearranged version of the same equation in EPRM. In these instances, both the blue *NCEES Handbook* equation and the EPRM equation will be shown along with useful commentary and an explanation of the differences.

Perhaps the most valuable feature of this book are the notes interspersed throughout. You'll find important equation derivations and explanations of how concepts align between the *NCEES Handbook* and EPRM that only an experienced electrical engineer can point out. The notes contain explanations of terms used in equations, units used for certain variables, underlying assumptions, and even simplified versions of equations you can apply to the exam. Where applicable, the notes also include helpful tips and information that elaborate on commonly confusing concepts. These tips will be priceless to you on exam day.

5. DEFINITIONS

At the end of each chapter, you will find a list of essential terms pertaining to the equations and concepts in the chapter. Occasionally, a term is defined in the notes for its relevant equation or concept; otherwise, terms will appear in the **Definitions** section at the end of the chapter.

6. NOMENCLATURE

At the end of each chapter, you will also find a **Nomenclature** section. This section contains a comprehensive list of the variables, symbols, and subscripts for all equations presented in the chapter.

Codes and Standards

The information used to write this book is based on the exam specifications current at the time of publication. However, just as state and local agencies do not always adopt codes, standards, and regulations as soon as they are issued, the PE exam is not always based on the most current codes. It is likely that the codes that are most current, the codes that you use in practice, and the codes that are the basis of your exam are all different. However, differences between code editions typically minimally affect the technical accuracy of this book, and the methodology presented remains valid. For more information about the variety of codes related to electrical engineering, refer to the following organizations and their websites.

- American National Standards Institute (ansi.org)

- Electronic Components Industry Association (ecianow.org)

- Federal Communications Commission (fcc.gov)

- Institute of Electrical and Electronics Engineers (ieee.org)

- International Organization for Standardization (iso.org)

- International Society of Automation (isa.org)

- National Electrical Manufacturers Association (nema.org)

- National Fire Protection Association (nfpa.org)

The PPI website (**ppi2pass.com**) provides the dates and editions of the codes, standards, and regulations on which the PE exams are based. It is your responsibility to check the NCEES website (ncees.org) and find out which codes are relevant to your exam.

Electronic versions of the following codes and standards will be provided in their entirety on exam day.

- ANSI C2-2017: *2017 National Electrical Safety Code* (NESC)

- NFPA 30B-2015: *Code for the Manufacture and Storage of Aerosol Products*

- NFPA 70-2017: *National Electrical Code* (NEC)

- NFPA 70E-2018: *Standard for Electrical Safety in the Workplace*

- NFPA 497-2017: *Recommended Practice for the Classification of Flammable Liquids, Gases, or Vapors and of Hazardous (Classified) Locations for Electrical Installations in Chemical Process Areas*

- NFPA 499-2017: *Recommended Practice for the Classification of Combustible Dusts and of Hazardous (Classified) Locations for Electrical Installations in Chemical Process Areas*

These are listed with bibliographic information in the following section.

CODES AND STANDARDS USED IN THIS BOOK

47 CFR 73: *Code of Federal Regulations*, "Title 47—Telecommunication, Part 73—Radio Broadcast Services," 2018. Office of the Federal Register National Archives and Records Administration, Washington, DC.

IEEE/ASTM SI 10: *American National Standard for Metric Practice*, 2016. ASTM International, West Conshohocken, PA.

IEEE Std. 141 (IEEE Red Book): *IEEE Recommended Practice for Electric Power Distribution for Industrial Plants*, 1993. The Institute of Electrical and Electronics Engineers, Inc., New York, NY.

IEEE Std. 142 (IEEE Green Book): *IEEE Recommended Practice for Grounding of Industrial and Commercial Power Systems*, 2007.

IEEE Std. 241 (IEEE Gray Book): *IEEE Recommended Practice for Electric Power Systems in Commercial Buildings*, 1990.

IEEE Std. 242 (IEEE Buff Book): *IEEE Recommended Practice for Protection and Coordination of Industrial and Commercial Power Systems*, 2001.

IEEE Std. 399 (IEEE Brown Book): *IEEE Recommended Practice for Industrial and Commercial Power Systems Analysis*, 1997.

IEEE Std. 446 (IEEE Orange Book): *IEEE Recommended Practice for Emergency and Standby Power Systems for Industrial and Commercial Applications*, 1995.

IEEE Std. 493 (IEEE Gold Book): *IEEE Recommended Practice for the Design of Reliable Industrial and Commercial Power Systems*, 2007.

IEEE Std. 551 (IEEE Violet Book): *IEEE Recommended Practice for Calculating AC Short-Circuit Currents in Industrial and Commercial Power Systems*, 2006.

IEEE Std. 602 (IEEE White Book): *IEEE Recommended Practice for Electric Systems in Health Care Facilities*, 2007.

IEEE Std. 739 (IEEE Bronze Book): *IEEE Recommended Practice for Energy Management in Industrial and Commercial Facilities*, 1995.

IEEE Std. 902 (IEEE Yellow Book): *IEEE Guide for Maintenance, Operation, and Safety of Industrial and Commercial Power Systems*, 1998.

IEEE Std. 1015 (IEEE Blue Book): *IEEE Recommended Practice for Applying Low-Voltage Circuit Breakers Used in Industrial and Commercial Power Systems*, 2006.

IEEE Std. 1100 (IEEE Emerald Book): *IEEE Recommended Practice for Powering and Grounding Electronic Equipment*, 2005.

NEC (NFPA 70): *National Electrical Code*, 2017. National Fire Protection Association, Quincy, MA.

NESC (ANSI C2): *2017 National Electrical Safety Code*, 2017. The Institute of Electrical and Electronics Engineers, Inc., New York, NY.

NFPA 30: *Flammable and Combustible Liquids Code*, 2018. National Fire Protection Association, Quincy, MA.

NFPA 30B: *Code for the Manufacture and Storage of Aerosol Products*, 2015.

NFPA 70E: *Standard for Electrical Safety in the Workplace*, 2018.

NFPA 497: *Recommended Practice for the Classification of Flammable Liquids, Gases, or Vapors and of Hazardous (Classified) Locations for Electrical Installations in Chemical Process Areas*, 2017.

NFPA 499: *Recommended Practice for the Classification of Combustible Dusts and of Hazardous (Classified) Locations for Electrical Installations in Chemical Process Areas*, 2017.

REFERENCES

Anthony, Michael A. *NEC Answers*. New York, NY: McGraw-Hill. (*National Electrical Code* example applications textbook.)

Bronzino, Joseph D. *The Biomedical Engineering Handbook*. Boca Raton, FL: CRC Press. (Electrical and electronics handbook.)

Chemical Rubber Company. *CRC Standard Mathematical Tables and Formulae*. Boca Raton, FL: CRC Press. (General engineering reference.)

Croft, Terrell, and Wilford I. Summers. *American Electricians' Handbook*. New York, NY: McGraw-Hill. (Power handbook.)

Earley, Mark W., et al. *National Electrical Code Handbook*, 2017 ed. Quincy, MA: National Fire Protection Association. (Power handbook.)

Fink, Donald G., and H. Wayne Beaty. *Standard Handbook for Electrical Engineers*. New York, NY: McGraw-Hill. (Power and electrical and electronics handbook.)

Grainger, John J., and William D. Stevenson, Jr. *Power System Analysis*. New York, NY: McGraw-Hill. (Power textbook.)

Horowitz, Stanley H., and Arun G. Phadke. *Power System Relaying*. Chichester, West Sussex: John Wiley & Sons, Ltd. (Power protection textbook.)

Huray, Paul G. *Maxwell's Equations*. Hoboken, NJ: John Wiley & Sons, Inc. (Power and electrical and electronics textbook.)

Jaeger, Richard C., and Travis Blalock. *Microelectronic Circuit Design*. New York, NY: McGraw-Hill Education. (Electronic fundamentals textbook.)

Lee, William C.Y. *Wireless and Cellular Telecommunications*. New York, NY: McGraw-Hill. (Electrical and electronics handbook.)

Marne, David J. *National Electrical Safety Code (NESC) 2017 Handbook*. New York, NY: McGraw-Hill Professional. (Power handbook.)

McMillan, Gregory K., and Douglas Considine. *Process/Industrial Instruments and Controls Handbook*. New York, NY: McGraw-Hill Professional. (Power and electrical and electronics handbook.)

Millman, Jacob, and Arvin Grabel. *Microelectronics*. New York, NY: McGraw-Hill. (Electronic fundamentals textbook.)

Mitra, Sanjit K. *An Introduction to Digital and Analog Integrated Circuits and Applications*. New York, NY: Harper & Row. (Digital circuit fundamentals textbook.)

Parker, Sybil P., ed. *McGraw-Hill Dictionary of Scientific and Technical Terms*. New York, NY: McGraw-Hill. (General engineering reference.)

Plonus, Martin A. *Applied Electromagnetics*. New York, NY: McGraw-Hill. (Electromagnetic theory textbook.)

Rea, Mark S., ed. *The IESNA Lighting Handbook: Reference & Applications*. New York, NY: Illuminating Engineering Society of North America. (Power handbook.)

Shackelford, James F., and William Alexander, eds. *CRC Materials Science and Engineering Handbook*. Boca Raton, FL: CRC Press. (General engineering handbook.)

Van Valkenburg, M.E., and B.K. Kinariwala. *Linear Circuits*. Englewood Cliffs, NJ: Prentice-Hall. (AC/DC fundamentals textbook.)

Wildi, Theodore, and Perry R. McNeill. *Electrical Power Technology*. New York, NY: John Wiley & Sons. (Power theory and application textbook.)

Measurement and Instrumentation

Exam specification 1.A, Measurement and Instrumentation, makes up between 5% and 8% of the PE Electrical Power exam (between 4 and 6 questions out of 80).

The organization of this chapter follows the order of knowledge areas given by the NCEES for this exam specification. Each knowledge area is covered in the following numbered sections.

Content in blue refers to the *NCEES Handbook.*

Content in red is additional essential information.

1. INSTRUMENT TRANSFORMERS

Instrument transformers isolate and transform voltage or current levels, and are designed to reproduce the voltage or current of the primary circuit in the secondary circuit, leaving the phase relationship and waveform unchanged. Instrumentation is used to measure the power angle, impedance angle, current angle, and most often, the power factor angle.

Knowledge Area Overview

Key concepts: These key concepts are important for answering exam questions in knowledge area 1.A.1, Instrument Transformers.

- power factor

- two-wattmeter method

- function and properties of instrument transformers in protection circuits

- current transformers (CTs) in wye and delta connections

- conditions leading to current transformer saturation

- current transformer ratio (CTR)

- minimum current needed to activate protection circuits

- appropriate tap settings for relays in a protection circuit

- measurement of electrical properties using the following circuits or instruments

  ∘ DC voltmeter

  ∘ d'Arsonval movement

  ∘ bridge circuit

  ∘ permanent magnet moving coil

PE Power Reference Manual **(EPRM):** Study these sections in EPRM that either relate directly to this knowledge area or provide background information.

- Section 17.22: Complex Power and the Power Triangle

- Section 27.10: Terminal Marking and Polarity

- Section 27.13: General Transformer Classifications

- Section 31.8: Protection System Elements

- Section 31.9: Transformers

- Section 31.18: IEEE Buff Book

- Section 41.1: Fundamentals

- Section 41.9: Power Measurements

- Section 41.10: Power Factor Measurements

- Section 41.11: Electronic Instrumentation

- Section 41.14: Measurement Standards and Conventions

NCEES Handbook: To prepare for this knowledge area, familiarize yourself with these sections in the *Handbook.*

- Complex Power

- Complex Power Triangle (Inductive Load)

The following equations and figures are relevant for knowledge area 1.A.1, Instrument Transformers.

Power Factor

Handbook: Complex Power

EPRM: Sec. 41.10

$$\text{pf} = \cos\theta$$

The power factor is often one of the items measured.

The *power factor*, pf, is given by the ratio of the true power, P, to the apparent power, S.

$$\text{pf} = \frac{P}{S} \qquad 41.13$$

Complex Power

Handbook: Complex Power

EPRM: Sec. 41.14

$$\mathbf{S} = \mathbf{VI}^* = P + jQ$$

The angle associated with $\mathbf{S}$, and more specifically with the difference between $\mathbf{P}$ and $\mathbf{S}$, is the power angle, ϕ.

This is the same as the overall impedance angle, ϕ.

The magnitude of the overall impedance angle is equal to the power factor angle, which is usually shown as ϕ as well.

$$\text{pf} = \cos\phi = \frac{\mathbf{P}}{|\mathbf{S}|} \qquad 41.25$$

As mentioned, the power factor angle is labeled as lagging or leading, though it could be labeled as positive or negative. A useful memory aid is "ELI the ICE man," meaning voltage, E, leads current, I, in an inductive, L, circuit (ELI), and current, I, leads voltage, E, in a capacitive, C, circuit (ICE).

Two-Wattmeter Method

EPRM: Sec. 41.10

$$\tan\phi = \frac{\sqrt{3}\,(P_H - P_L)}{P_H + P_L} \qquad 41.14$$

Using the two-wattmeter method, the tangent of the power factor angle can be determined from the high and low power measured.

Power Triangle

Handbook: Complex Power Triangle (Inductive Load)

EPRM: Sec. 17.22, Sec. 41.14

A power triangle shows the relationships between apparent power, S, real power, P, and reactive power, Q, in an AC circuit.

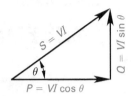

The power triangle is discussed in EPRM Sec. 17.22 and Sec. 41.14. If the voltage leads the current, the load is inductive and the power factor angle is positive. If the current leads the voltage, the load is capacitive and the power factor angle is negative.

The power factor angle is given as theta, θ, in the *Handbook* and phi, ϕ, in EPRM.

Figure 17.11 *Power Triangle*

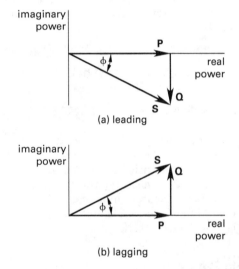

Current Transformers

EPRM: Sec. 31.9

Current transformers (CTs) are used as instrument transformers (also known as *metering transformers*) or as transducers for protection circuits (i.e., relaying transformers).

Figure 31.12 *Current Transformer*

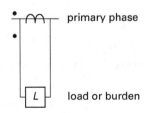

The primary of the CT is connected, in series, to the line and the secondary to the meter. The CT transforms the current to a lower value without changing the expected magnitude or the phase relationship. This is used for measuring the high value of current.

The *current transformer ratio* (CTR) is the ratio of the primary current to the secondary current.

$$\text{CTR} = \frac{I_p}{I_s} = \frac{I_{\text{pri}}}{I_{\text{sec}}}$$

The burden is the amount of power drawn from the circuit connecting the secondary terminals of instruments, usually given as apparent power in volt-amperes.

Tap Settings

EPRM: Sec. 27.10

Tappings are provided on a transformer winding for adjusting the number of turns on the transformer winding to maintain the output voltage within the desired limit.

Voltage Transformers

EPRM: Sec. 31.9

Voltage transformers (also known as *potential transformers*) are used for metering or relaying. The voltage transformer transforms the voltage to a lower value without changing the expected magnitude and phase relationship.

Figure 31.13 *Voltage Transformer*

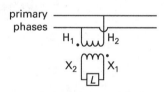

(a) voltage transformer

(b) additive polarity

(c) subtractive polarity

The primary of the voltage transformer is connected, in parallel, to the line and the secondary to the meter.

The transformers can have either additive or subtractive polarities. In EPRM Fig. 31.13, the H connection represents the primary phase and the X connection represents the secondary phase. If the X terminal is connected diagonally across the H terminal, the transformer has additive polarity. If the X terminal is connected directly across the H terminal, the transformer has subtractive polarity.

The change in voltage output from no-load to full-load is called the voltage regulation.

2. INSULATION TESTING

Insulation testing is conducted to ensure equipment can operate properly and personnel are protected. A successful insulation test will yield a high resistance value. Generally, 1 MΩ indicates satisfactory resistance, though this value varies with the applicable standard.

Knowledge Area Overview

Key concepts: These key concepts are important for answering exam questions in knowledge area 1.A.2, Insulation Testing.

- testing of insulation using instruments

- absorption current measurement

- effect of temperature on insulation resistance

- temperature coefficient for thermosetting and thermoplastic insulation systems

- insulation polarization index

PE Power Reference Manual (**EPRM**): Study these sections in EPRM that either relate directly to this knowledge area or provide background information.

- Section 26.6: Underground Distribution

- Section 27.3: Transformer Capacity

- Section 27.14: Buck Transformers/Boost Transformers/Autotransformers

- Section 31.2: Power System Grounding

- Section 31.7: Relay Speed

- Section 32.13: Service Factor

- Section 32.14: Motor Classifications

- Section 33.3: Service Factor

- Section 33.4: Motor Classifications

- Section 34.13: Schottky Diodes

- Section 38.2: Concepts and Definitions

- Section 38.5: Protective Devices
- Section 41.12: Insulation and Ground Testing
- Section 43.7: Shock Protection
- Section 44.5: Wiring and Protection: Grounded Conductors
- Section 44.11: Wiring Methods and Materials
- Section 45.1: Overview
- Section 45.6: NESC Part 1

NCEES Handbook: To prepare for this knowledge area, familiarize yourself with these sections in the *Handbook*.

- Insulation Resistance Theory
- Characteristics of the Measured Direct Current
- Insulation Testing: Minimum Insulation Resistance
- Effect of Temperature
- Insulation Polarization Index (PI)

The following equations are relevant for knowledge area 1.A.2, Insulation Testing.

Absorption Current

Handbook: Insulation Resistance Theory; Characteristics of the Measured Direct Current

$$I_A = Kt^{-n}$$

The absorption equation from the *Handbook* consists of factors K and n, which are dependent upon the voltage, capacitance, and insulation system used. These are measured values and cannot be calculated. The actual values would have to be provided or listed in a table.

The formula is the best fit of data from a semi-log graph of test results.

EPRM Eq. 41.15 can be used to calculate insulation resistance.

$$R = \frac{t}{2.303 C \log_{10}\left(\dfrac{d_1}{d_2}\right)} \quad \text{[in the units of M}\Omega\text{]} \quad \textit{41.15}$$

The values used in EPRM Eq. 41.15 can be taken from an electronic megger using a *leakage method* or *loss of charge method*. C is the capacitance and both d_1 and d_2 represent the deflections that occur as a switch is open and shut over time, t. This is very similar to the measurements over time discussed in the *Handbook*.

Minimum Insulation Resistance

Handbook: Insulation Testing: Minimum Insulation Resistance

The recommended minimum resistance, $R_{I,1\,min}$, in megohms is the observed insulation resistance, corrected to 40°C, obtained by applying a constant direct voltage to the entire winding for 1 minute. All of the following conditions and levels of testing come from IEEE Std. 43.

Before 1970, the following minimum resistance applies.

$$R_{I,1min} = kV_{rated} + 1 \quad \text{[most windings before 1970]}$$

This is the minimum 1-minute insulation testing resistance in megohms for most windings. For example, a 1 kV rated winding would be tested to 2 MΩ.

After 1970, the following minimum resistance applies.

$$R_{I,1min} = 100 \quad \text{[most windings after 1970]}$$

This is the minimum 1-minute insulation testing resistance in megohms for AC windings built after 1970. For example, all such windings would be tested to 100 MΩ.

For most machines with random-wound stator coils and form-wound coils rated below 1 kV and DC armatures, the following minimum resistance applies.

$$R_{I,1min} = 5$$

This is the minimum 1-minute insulation testing resistance in megohms for wound coil windings. For example, all such windings would be tested to 5 MΩ.

Effect of Temperature on Insulation Resistance

Handbook: Effect of Temperature

$$R_{I,40°C} = k_{T°C} R_{I,T°C}$$

Insulation resistance is affected by the increase in temperature.

This equation corrects a resistance at a given temperature to the resistance at the standard 40°C using the resistance temperature coefficient, k, for the given material.

The coefficient can be calculated based on the type of insulation.

Coefficient *k* for Thermosetting Insulation Systems

Handbook: Effect of Temperature

$$k_{T^\circ C} = \exp\left(-4230\left(\frac{1}{T+273} - \frac{1}{313}\right)\right)$$
$$[\text{for } 40^\circ C \le T < 85^\circ C]$$

$$k_{T^\circ C} = \exp\left(-1245\left(\frac{1}{T+273} - \frac{1}{313}\right)\right)$$
$$[\text{for } 10^\circ C < T < 40^\circ C]$$

The preceding equations are for thermosetting-type insulation materials.

A thermosetting material is a substance that sets permanently when heated.

Coefficient *k* for Thermoplastic Insulation Systems

Handbook: Effect of Temperature

$$k_{T^\circ C} = 0.5^{(40-T)/10}$$

The preceding equation is for thermoplastic-type insulation materials.

A thermoplastic material is a substance that becomes plastic upon heating and solid once cooled.

Polarization Index

Handbook: Insulation Polarization Index (PI)

$$PI = \frac{R_{I,10\,\text{min}}}{R_{I,1\,\text{min}}}$$

The polarization index (PI) is a measure of insulation resistance over time used to determine insulation health.

It is the ratio of the 10-minute resistance value, $R_{I,10\,\text{min}}$, to the 1-minute resistance value, $R_{I,1\,\text{min}}$.

Generally, a PI of 2 is considered a minimum acceptable value.

3. GROUND RESISTANCE TESTING

The term *ground testing* is often used interchangeably to mean both equipment insulation checks and earth ground checks. Earth ground checks are performed less often but are just as critical, if not more so, than other ground testing. Ground testing determines if a conducting connection exists between an electrical circuit or equipment and the ground or some conducting body.

Knowledge Area Overview

Key concepts: These key concepts are important for answering exam questions in knowledge area 1.A.3, Ground Resistance Testing.

- ground resistance testing using various methods

  - Wenner method

  - Schlumberger method

  - variation of depth method

- appropriate methods for lowering resistance

PE Power Reference Manual (**EPRM**): Study these sections in EPRM that either relate directly to this knowledge area or provide background information.

- Section 16.3: Resistance

- Section 41.12: Insulation and Ground Testing

- Section 41.13: Earth Grounding Electrode Systems

NCEES Handbook: To prepare for this knowledge area, familiarize yourself with this section in the *Handbook*.

- Ground Resistance Testing

The following equations and figures are relevant for knowledge area 1.A.3, Ground Resistance Testing.

Equally Spaced 4-Pin Method, or the Wenner Method

Handbook: Ground Resistance Testing

EPRM: Sec. 41.13

In a Wenner array, four electrodes are equally spaced in line.

Figure 41.14 Equally Spaced Four-Pin Method (Wenner Method)

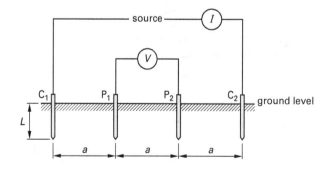

The measured ground resistance is

$$R = \frac{V}{I}$$

The equation used in the *Handbook* is merely a rearrangement of Ohm's law as seen in EPRM Eq. 7.32. Ohm's law itself is not in the *Handbook*.

The apparent resistivity is determined using the following two equations.

$$\rho = \frac{4\pi a R}{1 + \dfrac{2a}{\sqrt{a^2 + 4L^2}} - \dfrac{a}{\sqrt{a^2 + L^2}}}$$

For $L \leq 0.1a$, ignore L.

$$\rho = 2\pi a R$$

The apparent resistivity of the soil (earth) is shown in the second equation.

If the electrodes are driven to a depth, L, that is less than one-tenth of the distance between the electrodes (a), then the resistivity can be simplified to the second equation.

The resistance, R, is usually limited to 25 Ω (by the NEC) or less depending upon the standard used.

While the resistivity of the soil can be calculated, the initial ground rod design parameters are usually provided using known or estimated values (from surveys) of soil resistivity.

In general terms, the resistivity, length, and area are combined to obtain the resistance, as shown in EPRM Eq. 16.1.

$$R = \frac{\rho l}{A} \qquad\qquad \textbf{16.1}$$

Unequally Spaced 4-Pin Method, or the Schlumberger Method

Handbook: Ground Resistance Testing

EPRM: Sec. 41.13

In the Schlumberger array, the two outer electrodes are placed at a larger distance than that of the two inner electrodes around a common point.

Figure 41.15 *Unequally Spaced Four-Pin Method (Schlumberger Method)*

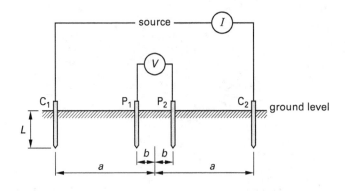

The measured ground resistance is

$$R = \frac{V}{I}$$

The equation used in the *Handbook* is merely a rearrangement of Ohm's law as seen in EPRM Eq. 7.32. Ohm's law itself is not in the *Handbook*.

The apparent resistivity is determined using the following two equations.

$$\rho = \frac{\pi a(a + 2b) R}{2b}$$

$$\rho = \frac{\pi a^2 R}{2b} \qquad \text{[for } a \gg 2b\text{]}$$

The apparent resistivity of the soil (earth) is shown in the first equation.

If the outer probe spacing (a) is much greater than the inner probe spacing (b), that is $a \gg 2b$, then the resistivity can be simplified to the second equation.

The resistance, R, is usually limited to 25 Ω (by the NEC) or less depending upon the standard used.

While the resistivity of the soil can be calculated, the initial ground rod design parameters are usually provided using known or estimated values (from surveys) of soil resistivity.

Variation of Depth Method, or the Driven Rod Method

Handbook: Ground Resistance Testing

EPRM: Sec. 41.13

For this method of substation grounding, the depth of the driven rod located in the soil is varied.

Figure 41.16 Driven Rod Method

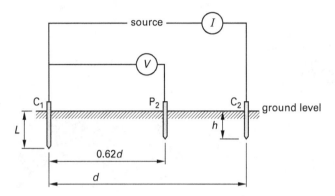

The ground resistance of the buried rod (C_1 in the diagram) is

$$R = \frac{V}{I}$$

The equation used in the *Handbook* is merely a rearrangement of Ohm's law as seen in EPRM Eq. 7.32. Ohm's law itself is not in the *Handbook*.

The approximate ground resistance of the buried rod is

$$R = \left(\frac{\rho}{2\pi L}\right)\left(\left(\ln\frac{4L}{\left(\frac{d}{2}\right)}\right) - 1\right)$$

The ground resistivity is

$$\rho = \frac{2\pi R L}{\left(\ln\frac{4L}{\left(\frac{d}{2}\right)}\right) - 1}$$

The depth of the driven rod is L. (It is l in the *Handbook*. L is used here for clarity.)

d is the distance from the electrode (rod) to the current probe—the place where the current is to be injected. The voltage probe is placed at $0.62d$. This is often called the 62% rule.

The resistance, R, is usually limited to 25 Ω (by the NEC) or less depending upon the standard used.

While the resistivity of the soil can be calculated, the initial ground rod design parameters are usually provided using known or estimated values (from surveys) of soil resistivity.

4. DEFINITIONS

absorption current: The current that is required to polarize the insulating medium; also called polarization current.

apparent resistivity: A measurement of resistivity calculated as the product of the measured resistance, R, and a geomagnetic factor.

burden: The amount of power drawn from a circuit connecting the secondary terminals of an instrument transformer, usually given in units of volt-ampere (VA).

current transformer: Transforms the current to a lower value without changing the expected magnitude or the phase relationship.

current transformer ratio: The ratio of the primary current to the secondary current.

instrument transformer: Transformers that isolate and transform voltage or current levels of the primary circuit in the secondary circuit, with the phase relationship and waveform effectively unchanged.

minimum insulation resistance: The observed insulation resistance, corrected to 40°C, obtained by applying a constant direct voltage to the entire winding for 1 minute.

polarization index: A measure of insulation resistance over time used to determine insulation health. It is the ratio of the 10-minute resistance value to the 1-minute resistance value.

power factor: The ratio of the true power to the apparent power.

soil resistivity: A measure of how much soil resists the conduction of current.

substation: A part of an electrical generation, transmission, and distribution system. Substations transform voltage from high to low, or the reverse, or perform any of several other important functions.

thermoplastic material: Substance that becomes plastic upon heating and solid once cooled.

thermosetting material: Substance that sets permanently when heated.

voltage transformer: Transforms the voltage to a lower value without changing the expected magnitude or the phase relationship; also known as potential transformers.

5. NOMENCLATURE

a	probe spacing	m
a, A	area	m^2
b	inner probe spacing	m
C	capacitance	F
CTR	current transformer ratio	–
d	deflection	varies
d	diameter	m
H	height, depth	m
I	current	A
k	temperature coefficient	–
K	proportionality constant	–
l, L	length	m
n	constant	–
P	real power	W
P	true power	W
pf	power factor	–
PI	polarization index	–
Q	reactive power	VAR
R	resistance	Ω
S	apparent power	VA
T	temperature	Celsius
t	time	sec
V	voltage	V

Symbols

θ	power angle	radians
ϕ	impedance angle, power angle	radians
ρ	resistivity	$\Omega \cdot$m

Subscripts

A	absorption
f	fault (current)
H	high
I	insulation
L	low
p	primary
pri	primary
s	secondary
sec	secondary

2 Applications

Exam specification 1.B, Applications, makes up between 9% and 14% of the PE Electrical Power exam (between 7 and 11 questions out of 80).

The organization of this chapter follows the order of knowledge areas given by the NCEES for this exam specification. Each knowledge area is covered in the following numbered sections.

Content in blue refers to the *NCEES Handbook*.

Content in red is additional essential information.

1. LIGHTNING PROTECTION

Design protection for lightning strikes is a coordinated affair using shielding, grounding, and surge arresters. Additionally, switching resistors, breaker timing, and surge capacitors protect against the resultant electrical transients.

Knowledge Area Overview

Key concepts: These key concepts are important for answering exam questions in knowledge area 1.B.1, Lightning Protection.

- lightning protection systems and processes
- properties of lightning
- basic lightning impulse insulation level (BIL)
- lightning strike frequency

PE Power Reference Manual **(EPRM):** Study these sections in EPRM that either relate directly to this knowledge area or provide background information.

- Section 8.4: Noise
- Section 23.9: Energy Management

- Section 23.10: Power Quality
- Section 31.2: Power System Grounding
- Chapter 38: Lightning Protection and Grounding
- Section 44.10: Wiring and Protection: Grounding
- Section 44.20: Hazardous Area Classification: Aerosol Products (NFPA 30B)
- Section 45.5: Grounding
- Section 45.6: NESC Part 1
- Section 45.7: NESC Part 2

NCEES Handbook: To prepare for this knowledge area, familiarize yourself with these sections and tables in the *Handbook*.

- Lightning Protection
- Location Factor, C_D
- Determination of Construction Coefficient, C_2
- Determination of Structure Contents Coefficient, C_3
- Determination of Structure Occupancy Coefficient, C_4
- Determination of Lightning Consequence Coefficient, C_5

The following equations, figures, and tables are relevant for knowledge area 1.B.1, Lightning Protection.

Annual Lightning Strike Frequency to a Structure

Handbook: Lightning Protection

$$N_D = N_G A_D C_D \text{ events/year}$$

N_G represents the annual lightning ground (subscript G) flash density and is an empirically determined value located in NFPA 780, a National Fire Protection Association publication. The units are flashes/unit area/year. Area may be in either SI units or U.S. customary units.

A_D is the structure's equivalent collection area—abbreviated as A_e in NFPA 780. The formula for equivalent collection area (shown in the next section) is dependent upon the dimensions of the structure.

C_D is a location factor based on surrounding foliage, structures, and topology. These factors are provided in *Handbook* table Location Factor, C_D. In NFPA 780, the subscript D is a location identifier, which in this case is 1, the environmental coefficient.

Overall then, N_D is the strike frequency density (subscript D) to the object or structure in question. The units are events/year.

Equivalent Collection Area of a Rectangular Structure

Handbook: Lightning Protection

$$A_D = LW + 6H(L + W) + \pi 9H^2$$

This equation depends solely on a structure's dimensions, but note that it is limited to rectangular structures only.

Tolerable Lightning Frequency

Handbook: Lightning Protection

$$N_c = \frac{1.5 \times 10^{-3}}{C} \text{ events/year}$$

The term C is the product of the structural coefficients C_2, C_3, C_4, and C_5 provided in *Handbook* tables Determination of Construction Coefficient, C_2; Determination of Structure Contents Coefficient, C_3; Determination of Structure Occupancy Coefficient, C_4; and Determination of Lightning Consequence Coefficient, C_5.

The tolerable lightning frequency is related to the risk of damage to property and personnel. The step voltage and contact voltage are limited to minimize risk to personnel. (These topics are covered in this book in knowledge area 1.A.2, Insulation Testing, and Chapter 3, Codes and Standards.)

Because of the risk to personnel, a protection scheme must contain all the grounding systems shown in EPRM Fig. 38.4.

Figure 38.4 *Grounding Systems*

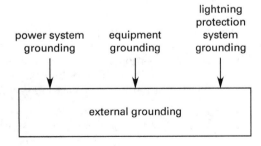

Standard Lightning Impulse

EPRM: Sec. 38.2

The standard lightning impulse voltage has a front time of 1.2 μs and a time to half-value of 50 μs. These measurements are obtained from field observation.

Figure 38.2 *Standard Lightning Impulse*

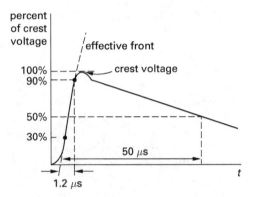

2. SURGE PROTECTION

Surge protection minimizes transients to protect equipment.

Knowledge Area Overview

Key concepts: These key concepts are important for answering exam questions in knowledge area 1.B.2, Surge Protection.

- troubleshooting techniques and preventative tips
- protective margin
- protection quality index
- surge protection terms
- power quality problems and their effects on different devices

PE Power Reference Manual **(EPRM):** Study these sections in EPRM that either relate directly to this knowledge area or provide background information.

- Section 15.4: Characteristic Impedance
- Section 23.7: Parallel Operation
- Section 23.10: Power Quality
- Section 26.3: Common-Neutral System
- Section 26.12: IEEE Red Book
- Section 27.14: Buck Transformers/Boost Transformers/Autotransformers
- Section 28.1: Fundamentals

- Section 38.2: Concepts and Definitions
- Section 38.5: Protective Devices
- Section 44.3: General
- Section 44.4: Wiring and Protection
- Section 44.10: Wiring and Protection: Grounding
- Section 44.20: Hazardous Area Classification: Aerosol Products (NFPA 30B)
- Section 45.5: Grounding
- Section 45.6: NESC Part 1
- Section 45.7: NESC Part 2

Codes and standards: These are the most important sections of the codes and standards for this knowledge area.

- NEC Art. 280: Surge Arresters, Over 1000 Volts
- NEC Art. 285: Surge-Protective Devices (SPDs), 1000 Volts or Less
- NESC Part 1, Sec. 19, Surge Arresters: Rules 190–193

NCEES Handbook: To prepare for this knowledge area, familiarize yourself with this section in the *Handbook*.

- Transmission Line Models

The following equations are relevant for knowledge area 1.B.2, Surge Protection.

Surge Impedance Loading

Handbook: Transmission Line Models

$$ \mathrm{SIL} = \frac{V_{\mathrm{line}}^2}{Z_s} $$

The SIL, or *surge impedance loading*, is the power expended on the impedance of the transmission line itself. It is a design parameter for surge protection.

The surge impedance is also known as the characteristic impedance, or intrinsic impedance, and depends upon the relative permeability, μ_r, and permittivity, ϵ_r, of the material.

The term shown for line surge impedance, Z_s, in the *Handbook* equation is equivalent to the term for surge impedance, Z, in EPRM Eq. 15.3. The surge impedance is

$$ Z = \sqrt{\frac{\mu}{\epsilon}} = \sqrt{\frac{\mu_r \mu_0}{\epsilon_r \epsilon_0}} = Z_0 \sqrt{\frac{\mu_r}{\epsilon_r}} \qquad 15.3 $$

Characteristic Impedance

Handbook: Transmission Line Models

EPRM: Sec. 28.1

$$ Z_c = \sqrt{\frac{z}{y}} $$

The characteristic impedance is the square root of the impedance per unit length divided by the admittance per unit length.

The terms shown in the *Handbook* equation are equivalent to those used in EPRM Eq. 28.1, where $Z_0 = Z_c$, $z = Z_l$, and $y = Y_l$.

Protection Quality Index

EPRM: Sec. 38.5

$$ \mathrm{PQI} = \frac{V_r}{V_p} \qquad 38.2 $$

The protection quality index is a measure of the effectiveness of a surge arrester.

V_r represents the reseal voltage level, which is the voltage level at which minimal current flows through the arrester. The ideal PQI value is 1.

V_p represents the voltage protection level provided by the surge arrester.

Protective Margin

EPRM: Sec. 38.5

$$ \mathrm{PM} = \frac{V_w - V_p}{V_p} \qquad 38.3 $$

The protective margin is a measure of the effectiveness of a surge arrester for a particular system.

V_w is the withstand voltage, which is the voltage that the device can withstand before damage will occur.

V_p is the voltage protection level provided by the surge arrester.

3. RELIABILITY

Increased reliability for utilities lowers maintenance and operations costs. Additionally, increased reliability lowers restoration costs, which occur following downtime. From a customer perspective, reliability of power may very well be the difference between life and death.

Knowledge Area Overview

Key concepts: These key concepts are important for answering exam questions in knowledge area 1.B.3, Reliability.

- mean time to failure (MTTF) for components

- reliability for parallel and series systems

- concepts of availability, downtime, and reliability

- availability of system components

- reliability for electrical power utilities

PE Power Reference Manual (EPRM): Study these sections in EPRM that either relate directly to this knowledge area or provide background information.

- Section 23.9: Energy Management

- Section 23.12: Application: Reliability

- Section 26.2: Classification of Distribution Systems

- Section 26.3: Common-Neutral System

- Section 28.9: Three-Phase Transmission

- Section 29.11: IEEE Brown Book

- Section 31.2: Power System Grounding

- Section 31.3: Power System Configurations

- Section 31.5: Relay Reliability

- Section 31.12: Solid-State Relays

- Section 31.19: IEEE Gold Book

- Section 31.20 IEEE Yellow Book

- Section 37.11: JFET Switches

- Section 40.7: IEEE Orange Book

- Section 44.18: Annexes

NCEES Handbook: To prepare for this knowledge area, familiarize yourself with this section in the *Handbook*.

- Reliability

The following equations are relevant for knowledge area 1.B.3, Reliability.

Independent Components Connected in Series

Handbook: Reliability

EPRM: Sec. 23.12

$$R(P_1, P_2, \cdots, P_n) = \prod_{i=1}^{n} P_i$$

EPRM Eq. 23.19 is a simplified way of showing that series component reliability is the product of the individual reliabilities.

$$p\{\Phi = 1\} = R_{\text{serial system}} = R_1 R_2 R_3 \cdots R_n \qquad \textbf{23.19}$$

Independent Components Connected in Parallel

Handbook: Reliability

EPRM: Sec. 23.12

$$R(P_1, P_2, \cdots, P_n) = 1 - \prod_{i=1}^{n}(1 - P_i)$$

The total reliability never exceeds the value of 1. EPRM Eq. 23.21 represents this starting value of 1 minus the paralleled component probabilities of failure (that is, one minus the reliability). The net result being the reliability of the paralleled system.

$$R = p\{\Phi = 1\}$$
$$= 1 - (1 - R_1)(1 - R_2)(1 - R_3) \cdots (1 - R_n) \qquad \textbf{23.21}$$

Mean Time to Failure

Handbook: Reliability

EPRM: Sec. 23.12

$$\text{MTTF} = \frac{1}{\lambda}$$

The mean time to failure (MTTF) is the expected time to failure for a non-repairable system. λ is the failure rate per unit of time. EPRM Eq. 23.14 is used to determine the failure rate and is a rearrangement of the *Handbook* equation shown.

Calculating Reliability

EPRM: Sec. 23.12

$$R\{t\} = e^{-\lambda t} = e^{-t/\text{MTTF}} \qquad \textbf{23.13}$$

$$R\{t\} = 1 - F(t) = 1 - (1 - e^{-\lambda t}) = e^{-\lambda t} \qquad \textbf{23.15}$$

The reliability $R(t)$ is 1 minus the failure rate, $F(t)$. The value of 1 being 100% success, no failures. The failure rate terminology is also used for the failures/hour and is given the symbol λ.

System Availability

Handbook: Reliability

$$A = \frac{\text{MTTF}}{\text{MTTF} + \text{MTTR}} = \frac{\text{MTTF}}{\text{MTBF}}$$

Availability is a measure of the ability of a piece of equipment to be operated if required. Reliability, by contrast, is the ability of a piece of equipment to perform its intended function for a specific timeframe without failure.

Mean time to repair (MTTR) is the average time to repair a piece of equipment.

Mean time between failures (MTBF) is the arithmetic mean (average) time between predicted failures of a system for a repairable system.

4. ILLUMINATION/LIGHTING AND ENERGY EFFICIENCY

The two common factors in illumination system design are efficiency and uniformity of the distribution. The term *illumination* is now deprecated and is not recommend for use. The preferred term now is *illuminance* with the symbol, E.

Knowledge Area Overview

Key concepts: These key concepts are important for answering exam questions in knowledge area 1.B.4, Illumination/Lighting and Energy Efficiency.

- luminous efficiency and illuminance

- properties of illuminance

- appropriate lighting levels using different lighting design methods

- cavity ratio

- coefficient of utilization

PE Power Reference Manual **(EPRM):** Study these sections in EPRM that either relate directly to this knowledge area or provide background information.

- Chapter 39: Illumination

- Chapter 44: National Electrical Code

NCEES Handbook: To prepare for this knowledge area, familiarize yourself with these sections in the *Handbook*.

- Illumination and Energy Efficiency: Nomenclature

- Illumination and Energy Efficiency: Calculations

The following equations are relevant for knowledge area 1.B.4, Illumination/Lighting and Energy Efficiency.

Lambert's Law

Handbook: Illumination and Energy Efficiency: Calculations

EPRM: Sec. 39.17

$$E = \frac{I}{D^2} \cos \theta$$

Lambert's law can be derived from the inverse square law.

$$E_2 = E_1 \cos \theta \qquad \textit{39.31}$$

Light Loss Factor, LLF

Handbook: Illumination and Energy Efficiency: Calculations

$$\text{LLF} = (\text{LLD})(\text{LDD})$$

The *light loss factor*, LLF, consists of two terms: the lumen lamp depreciation (LLD) and luminaire dirt depreciation (LDD). Both are recoverable losses in that they can be corrected.

When LLF is added to EPRM Eq. 39.58, the equation for the average initial expected illuminance, the result is the equation for determining the average minimum lighting level over time.

$$E_{\text{maintained}} = \frac{L_{\text{total}}(\text{CU})(\text{LLF})}{A_w} \qquad \textit{39.59}$$

LLF is usually taken from a tabulated value and includes additional recoverable losses such as room surface dirt depreciation (RSDD) and lamp burnout factor (LBO).

A designer must determine how many such losses should accounted for in a given design.

L is the luminance, and A is the area.

Initial Illumination Level, II

Handbook: Illumination and Energy Efficiency: Calculations

$$\text{II} = \frac{\text{MMI}}{(\text{CU})(\text{LDD})(\text{LLD})}$$

The *Handbook* equation is a rearrangement of EPRM Eq. 39.59.

$$E_{\text{maintained}} = \frac{L_{\text{total}}(\text{CU})(\text{LLF})}{A_w} \qquad \textit{39.59}$$

L_{total} in the EPRM equation is initial illumination level, II, in the *Handbook* equation.

The light loss factor, LLF, in the EPRM equation represents the two loss terms, LDD and LLD, in the *Handbook* equation.

$E_{\text{maintained}}$ in the EPRM equation is the equivalent of the MMI in the *Handbook* equation.

Different texts and authors use various abbreviations, subscripts, and terminology to represent identical concepts. Standards organizations exist in an effort to minimize such differences.

Luminous Flux

Handbook: Illumination and Energy Efficiency: Calculations

$$\Phi = \frac{(\text{II})\,WL}{\text{CU}}$$

II is the initial illuminance in foot-candles (fc).

CU is the coefficient of utilization, which is a factor representing how well the light is distributed in the area of concern.

W and L are the width and length of the area of concern.

Φ is the resulting luminous flux in the area in units of lumens.

Cavity Ratio, CR

Handbook: Illumination and Energy Efficiency: Calculations

$$\text{CR} = 2.5\left(\frac{\text{wall area of cavity}}{\text{area of cavity base}}\right)$$

EPRM: Sec. 39.35

$$\text{CR} = \frac{2.5 H_{xx} P}{A_{\text{base}}} \qquad \textit{39.62}$$

The cavity ratio formula shown is for a room of an irregular shape.

Such ratios are important in lighting design as they determine the amount of light exchanged between the cavities and thus the number and intensity of the lights required.

EPRM Eq. 39.62 represents all the cavity ratios for a room of an irregular shape depending upon the value substituted for H_{xx}. When replacing with H_{rc}, the result is the room cavity ratio. When replacing with H_{cc}, the result is the ceiling cavity ratio. When replacing with H_{fc}, the result is the floor cavity ratio.

H represents the height of the particular cavity.

Room Cavity Ratio

Handbook: Illumination and Energy Efficiency: Calculations

EPRM: Sec. 39.35

$$\text{RCR} = 5h_{\text{rc}}\left(\frac{L+W}{LW}\right)$$

The room cavity is the space in between the ceiling cavity and the floor cavity.

The formula in the *Handbook* is equivalent to EPRM Eq. 39.61.

EPRM Eq. 39.61 represents all the cavity ratios depending upon the value substituted for H_{xx}. When replacing with H_{rc}, the result is the room cavity ratio.

H represents the height of the particular cavity.

Ceiling Cavity Ratio

Handbook: Illumination and Energy Efficiency: Calculations

EPRM: Sec. 39.35

$$\text{CCR} = 5h_{\text{cc}}\left(\frac{L+W}{LW}\right)$$

The ceiling cavity is from the top of the windows to the ceiling.

The formula in the *Handbook* is for the room cavity ratio.

EPRM Eq. 39.61 represents all the cavity ratios depending upon the value substituted for H_{xx}. When replacing with H_{cc}, the result is the ceiling cavity ratio.

H represents the height of the particular cavity.

Floor Cavity Ratio

Handbook: Illumination and Energy Efficiency: Calculations

EPRM: Sec. 39.35

$$\text{FCR} = 5h_{\text{fc}}\left(\frac{L+W}{LW}\right)$$

The floor cavity is from the windowsill to the floor.

The formula in the *Handbook* is for the room cavity ratio.

EPRM Eq. 39.61 represents all the cavity ratios depending upon the value substituted for H_{xx}. When replacing with H_{fc}, the result is the floor cavity ratio.

H represents the height of the particular cavity.

Minimum Maintained Illumination Level

Handbook: Illumination and Energy Efficiency: Calculations

$$\text{MMI} = \frac{(\# \text{ luminaires})\left(\dfrac{\text{lamps}}{\text{luminaire}}\right) \times (\text{initial flux in lumens})(\text{LLF})(\text{CU})}{\text{area of working surface in ft}^2}$$

EPRM: Sec. 39.34

$$E_{\text{maintained}} = \frac{N_{\text{lights}}L_{\text{per light}}(\text{CU})(\text{LLF})}{A_w} \qquad \textit{39.60}$$

The two equations shown from the *Handbook* and EPRM are equivalent. Only the terminology and symbols differ. The number of lights, N_{lights}, in Eq. 39.60 includes the product of the number of luminaires and lamps per luminaire, while the term $L_{\text{per light}}$ is the flux for each luminaire.

Reflected Radiation Coefficient

Handbook: Illumination and Energy Efficiency: Calculations

$$\text{RRC} = \text{LC}_w + (\text{RPM})(\text{LC}_{cc} - \text{LC}_w)$$

Lighting design depends upon both direct light from sources and indirect or reflected lighting, such as from walls, ceilings, and floors.

The terms in the *Handbook* equation include the ceiling cavity luminance coefficient, LC_{cc}, the wall luminance coefficient, LC_w, and the room position multiplier, RPM, which accounts for the room dimensions.

Minimum Mounting Height

Handbook: Illumination and Energy Efficiency: Calculations

$$\text{MH}_{\min} = \frac{\text{max candlepower}}{1000} + \text{ND}$$

The mounting height, MH, of lighting depends on the candlepower of the lighting and a factor called the MH constant, ND, which is determined by the distribution of the light expected: short (5), medium (10), or long (15) per *Handbook* section Illumination and Energy Efficiency: Nomenclature.

Lighting standards also use type numbers for beam patterns.

Floodlight Requirement

Handbook: Illumination and Energy Efficiency: Calculations

$$\# \text{ floodlights} = \frac{(\text{illumination level})(\text{sq ft})}{(\text{beam lumens})\left(\dfrac{\text{lamps}}{\text{floodlight}}\right)(\text{LLF})(\text{CBU})}$$

The *Handbook* equation is a rearrangement of EPRM Eq. 39.60 with the addition of the coefficient of beam utilization (CBU), which factors in the efficiency of the light design—specifically, the light reaching a given area to that provided in the beam.

$$E_{\text{maintained}} = \frac{N_{\text{lights}}L_{\text{per light}}(\text{CU})(\text{LLF})}{A_w} \qquad \textit{39.60}$$

In EPRM Eq. 39.60, the term CU represents the coefficient of utilization, which is a measure of the efficiency of a luminaire in transferring luminous energy to a working plane in a particular area.

N_{lights} from the EPRM equation is equal to $N_{\text{lamps/floodlight}} \times N_{\text{floodlights}}$ in the *Handbook* equation.

5. DEMAND CALCULATIONS

Demand calculations are applicable to power plants, NEC loading calculations involving wiring, and lighting calculations.

Knowledge Area Overview

Key concepts: These key concepts are important for answering exam questions in knowledge area 1.B.5, Demand Calculations.

- annual load factor
- coincidence factor
- demand and utilization factor
- diversity factor
- loss factor
- plant factor

PE Power Reference Manual (**EPRM**): Study these sections in EPRM that either relate directly to this knowledge area or provide background information.

- Section 26.10: Smart Grid
- Section 30.23: Control System Types/Modes
- Section 40.4: Demand Calculations
- Section 44.8: Wiring and Protection: Branch Circuit, Feeder, and Service Calculations

- Section 44.9: Wiring and Protection: Overcurrent Protection

Codes and standards: These are the most important sections of the codes and standards for this knowledge area.

- NEC Art. 220: Branch-Circuit, Feeder, and Service Load Calculations

- NEC Art. 240: Overcurrent Protection

NCEES Handbook: To prepare for this knowledge area, familiarize yourself with this section in the *Handbook.*

- Demand Calculations

The following equations are relevant for knowledge area 1.B.5, Demand Calculations.

Demand Factor, DF

Handbook: Demand Calculations

EPRM: Sec. 40.4

$$DF = \frac{\text{maximum demand}}{\text{total connected demand}}$$

EPRM Eq. 40.5 gives an equivalent equation to the *Handbook* equation. D_{max} represents the maximum demand and $D_{connected}$ represents the connected demand.

The DF will always be less than or equal to one.

Utilization Factor, F_u

Handbook: Demand Calculations

EPRM: Sec. 40.4

$$F_u = \frac{\text{maximum demand}}{\text{rated system capacity}}$$

EPRM Eq. 40.6 gives an equivalent equation to the *Handbook* equation. D_{max} represents the maximum demand, and C_{rated} represents the rated system capacity.

The utilization factor will always be less than or equal to one.

Plant Factor, Pf, over a Period *T*

Handbook: Demand Calculations

EPRM: Sec. 40.4

$$Pf = \frac{\text{actual energy produced or served}}{(\text{plant rating})\,T}$$

The *Handbook* uses the abbreviation Pf to represent plant factor, but more often the abbreviation PF is used.

The *Handbook* equation and EPRM Eq. 40.7 are equivalent.

Load Factor, FLD

Handbook: Demand Calculations

EPRM: Sec. 40.4

$$FLD = \frac{\text{average load}}{\text{peak load}}$$

EPRM Eq. 40.8 is the equivalent of the *Handbook* equation.

Annual Load Factor

Handbook: Demand Calculations

EPRM: Sec. 40.4

$$\text{annual load factor} = \frac{\text{total annual energy}}{(\text{peak load})(8760)}$$

The annual load factor is the kWh used in a year divided by the peak load in kW that occurred in the year (8760 hours).

Electrical companies want a high load factor so that the utilization of the energy available is efficient.

EPRM Eq. 40.9 is the equivalent of the *Handbook* equation.

Diversity Factor, F_{div}

Handbook: Demand Calculations

EPRM: Sec. 40.4

$$F_{div} = \frac{\sum_{i=1}^{n} D_{i\text{-max}}}{D_g}$$

EPRM Eq. 40.11 and the *Handbook* equation are the same.

Coincidence Factor, F_c

Handbook: Demand Calculations

EPRM: Sec. 40.4

$$F_c = \frac{D_g}{\sum_{i=1}^{n} D_{i\text{-max}}} = \frac{1}{F_{div}}$$

The coincidence factor determines the likelihood that individual components are peaking at the same time.

The coincidence factor is the reciprocal of the diversity factor.

Load Diversity, LD

Handbook: Demand Calculations

EPRM: Sec. 40.4

$$LD = \sum_{i=1}^{n} D_{i\text{-max}} - D_g$$

The load diversity, LD, and the coincidence factor, F_c, are used for top-down forecasting for transmission load planning. The *Handbook* equation and EPRM Eq. 40.12 are identical.

Loss Factor, FLS

Handbook: Demand Calculations

EPRM: Sec 40.4

$$FLS = \frac{\text{average power loss}}{\text{power loss at peak load}}$$

FLS is a measure of the losses occurring in an electric distribution system.

Losses can be within the range of 3–4% in urban systems and 7–8% in rural areas.

EPRM Eq. 40.10 is the equivalent of the *Handbook* equation.

6. ENERGY MANAGEMENT

Energy management has numerous definitions but in the context of the exam is the optimization of energy consumption on both a short- and long-term basis and on small and large scales.

Knowledge Area Overview

Key concepts: These key concepts are important for answering exam questions in knowledge area 1.B.6, Energy Management.

- efficiency
- energy and power unit conversion
- doubling times
- per-unit growth rates

PE Power Reference Manual (**EPRM**): Study these sections in EPRM that either relate directly to this knowledge area or provide background information.

- Section 23.9: Energy Management
- Chapter 26: Power Distribution
- Chapter 40: Power System Management

NCEES Handbook: To prepare for this knowledge area, familiarize yourself with this section in the *Handbook*.

- Energy Management

The following equations are relevant for knowledge area 1.B.6, Energy Management.

Efficiency

Handbook: Energy Management

$$\% \text{ efficiency} = \frac{P_{\text{out}}}{P_{\text{in}}} \times 100\%$$

EPRM: Sec. 32.15

$$\eta = \frac{\text{output}}{\text{input}} = \frac{\text{output}}{\text{output} + \text{losses}}$$
$$= \frac{\text{input} - \text{losses}}{\text{input}}$$

32.20

The efficiency of any process is the desired power output, which does not include undesired power losses, divided by the input power.

η represents efficiency.

Energy

Handbook: Energy Management

$$E = \int_0^t P(t)\, dt$$

Power is the time rate of change of work. Thus, energy is the time integral of power.

Per Unit Growth Rate

EPRM: Sec. 40.2

$$E = E_0 e^{at}$$

40.3

This equation represents the energy at some time in the future knowing the current energy usage, E_0, and the per-unit growth rate, a.

Doubling Time

EPRM: Sec. 40.2

$$t_d = \frac{\ln 2}{a} = \frac{0.693}{a}$$

40.4

The doubling time is the time it takes for energy usage to double.

7. ENGINEERING ECONOMICS

An engineer requires a basic understanding of economics to determine the best value of various engineering solutions.

Knowledge Area Overview

Key concepts: These key concepts are important for answering exam questions in knowledge area 1.B.7, Engineering Economics.

- comparison of economic alternatives
- cash flow diagrams
- various methods of depreciation
- nomenclature used in engineering economics
- cash flow calculations
- conventions of basic engineering economics calculations

***PE Power Reference Manual* (EPRM):** Study these sections in EPRM that either relate directly to this knowledge area or provide background information.

- Chapter 47: Engineering Economic Analysis

NCEES Handbook: To prepare for this knowledge area, familiarize yourself with these sections and tables in the *Handbook.*

- Engineering Economics
- Engineering Economics: Nomenclature and Definitions
- Engineering Economics: Risk
- Non-Annual Compounding
- Breakeven Analysis
- Inflation
- Depreciation: Straight Line
- Depreciation: Modified Accelerated Cost Recovery System (MACRS)
- MACRS Factors
- Book Value
- Taxation
- Capitalized Costs
- Bonds
- Rate-of-Return
- Benefit-Cost Analysis
- Interest Rate Tables

The following equations and tables are relevant for knowledge area 1.B.7, Engineering Economics.

Economic Factor Conversions

Handbook: Engineering Economics

EPRM: Table 47.1

The complete list of factor names and corresponding formulas are available in both the *Handbook* and EPRM Chap. 47.

A method of remembering the notation is to interpret the factors algebraically. The (F/P) factor could be thought of as the fraction F/P. Algebraically, the (F/P) factor would be as shown in EPRM Eq. 47.6.

$$F = P\left(\frac{F}{P}\right) \qquad \textit{47.6}$$

Single Payment Compound Amount

Handbook: Engineering Economics

EPRM: Table 47.1

$$(F/P, i\%, n) = (1+i)^n$$

For the appropriate interest rate, i, and number of years, n, use this equation to calculate the future value, F, of a present worth, P, of money.

EPRM Eq. 47.2 is the same equation, but includes the economic factors F and P.

Single Payment Present Worth

Handbook: Engineering Economics

EPRM: Table 47.1

$$(P/F, i\%, n) = (1+i)^{-n}$$

For the appropriate interest rate, i, and number of years, n, this equation can calculate the present worth, P, of a future amount, F, of money.

Uniform Series Sinking Fund

Handbook: Engineering Economics

EPRM: Table 47.1

$$(A/F, i\%, n) = \frac{i}{(1+i)^n - 1}$$

This is the annual value, A, of the future worth, F, of money.

The A/F factor is called the sinking fund factor.

The cash flow equivalent factors for this equation can be tabulated using *Handbook* table Interest Rate Tables or using EPRM App. 47.B.

Capital Recovery

Handbook: Engineering Economics

EPRM: Table 47.1

$$(A/P, i\%, n) = \frac{i(1+i)^n}{(1+i)^n - 1}$$

This is the annual value, A, of the present worth, P, of the money.

The cash flow equivalent factors for this equation can be tabulated using *Handbook* table Interest Rate Tables or using EPRM App. 47.B.

Uniform Series Compound Amount

Handbook: Engineering Economics

EPRM: Table 47.1

$$(F/A, i\%, n) = \frac{(1+i)^n - 1}{i}$$

This is the future worth, F, of the annual value, A, of the money.

The cash flow equivalent factors for this equation can be tabulated using *Handbook* table Interest Rate Tables or using EPRM App. 47.B.

Uniform Series Present Worth

Handbook: Engineering Economics

EPRM: Table 47.1

$$(P/A, i\%, n) = \frac{(1+i)^n - 1}{i(1+i)^n}$$

This is the present worth, P, of the annual value, A, of the money.

The cash flow equivalent factors for this equation can be tabulated using *Handbook* table Interest Rate Tables or using EPRM App. 47.B.

Uniform Gradient Present Worth

Handbook: Engineering Economics

EPRM: Table 47.1

$$(G/P, i\%, n) = \frac{(1+i)^n - 1}{i^2(1+i)^n} - \frac{n}{i(1+i)^n}$$

This is the gradient cash flow, G, for the present worth, P, of the money.

The cash flow equivalent factors for this equation can be tabulated using *Handbook* table Interest Rate Tables or using EPRM App. 47.B.

Uniform Gradient Future Worth

Handbook: Engineering Economics

EPRM: Table 47.1

$$(G/F, i\%, n) = \frac{(1+i)^n - 1}{i^2} - \frac{n}{i}$$

This is the gradient cash flow, G, of the future worth, F, of the money.

The cash flow equivalent factors for this equation can be tabulated using *Handbook* table Interest Rate Tables or using EPRM App. 47.B.

Uniform Gradient Uniform Series

Handbook: Engineering Economics

EPRM: Table 47.1

$$(A/G, i\%, n) = \frac{1}{i} - \frac{n}{(1+i)^n - 1}$$

This is the annual value, A, of the gradient cash flow, G, of the money.

The cash flow equivalent factors for this equation can be tabulated using *Handbook* table Interest Rate Tables or using EPRM App. 47.B.

Non-Annual Compounding

Handbook: Non-Annual Compounding

$$i_e = \left(1 + \frac{r}{m}\right)^m - 1$$

r is the interest rate. m is the number of time intervals in a given year for which the compounding occurs.

Inflation

Handbook: Inflation

$$d = i + f + (i \times f)$$

EPRM: Sec. 47.56

$$i' = i + e + ie \qquad 47.73$$

In the *Handbook* equation, the term d is the interest rate period adjusted (deflated) for inflation rate f per interest rate period of i. It is used to calculate the present worth value, P, of the money.

In EPRM Eq. 47.73, the effective annual interest rate, i, is replaced with a value corrected for inflation, i'. The term e is the decimal inflation rate.

Assuming that inflation is constant, cash flows can be adjusted to $t = 0$ by dividing cash flows by the following term.

$$(1+e)^n$$

e is the decimal inflation rate and n is the year (or years) of the cash flow(s).

Straight-Line Depreciation

Handbook: Depreciation: Straight Line

EPRM: Sec 47.36

$$D_j = \frac{C - S_n}{n}$$

This is depreciation in year j, or D_j, which depends upon the cost, C, and salvage value in year n, S_n.

The *Handbook* equation is equivalent to EPRM Eq. 47.25.

Modified Accelerated Cost Recovery System (MACRS)

Handbook: Depreciation: Modified Accelerated Cost Recovery System (MACRS)

EPRM: Sec. 47.36

$$D_j = (\text{factor})\,C$$

The *Handbook* equation is equivalent to EPRM Eq. 47.33.

Book Value

Handbook: Book Value

EPRM: Sec. 47.38

$$BV = \text{initial cost} - \sum D_j$$

BV is the book value. The book value is determined by subtracting the depreciation at year j from the initial cost, which is represented by the term C in EPRM Eq. 47.43.

Benefit-Cost Analysis

Handbook: Benefit-Cost Analysis

$$B - C \geq 0, \text{ or } B/C \geq 1$$

EPRM: Sec. 47.26

$$\frac{B_2 - B_1}{C_2 - C_1} \geq 1 \text{ [alternative 2 superior]} \qquad 47.22$$

The *Handbook* equation compares benefits, B, to costs, C, of a single alternative.

The EPRM equation compares two alternatives. If the value is >1, then alternative 2 is superior, if not, alternative 1 is superior.

8. GROUNDING

Grounding in the context of this knowledge area refers to the process of providing a connection to earth for the protection of personnel, equipment, and prevention of lightning damage.

Knowledge Area Overview

Key concepts: These key concepts are important for answering exam questions in knowledge area 1.B.7, Grounding.

- substation configurations and their impact on system protection
- grounding of a power system
- NEC codes and NESC rules relating to grounding
- maximum allowable touch voltage and step voltage
- permissible body current limit
- effects of shock and burns
- behavior of protection system components

PE Power Reference Manual (**EPRM**): Study these sections in EPRM that either relate directly to this knowledge area or provide background information.

- Section 27.13: General Transformer Classifications
- Section 27.16: Zigzag Transformers
- Section 31.2: Power System Grounding
- Chapter 38: Lightning Protection and Grounding
- Section 43.7: Shock Protection
- Section 44.3: General
- Section 44.4: Wiring and Protection
- Section 44.5: Wiring and Protection: Grounded Conductors
- Section 44.10: Wiring and Protection: Grounding
- Section 45.5: Grounding

Codes and standards: This is the most important section of the codes and standards for this knowledge area.

- NEC Art. 250: Grounding and Bonding

NCEES Handbook: To prepare for this knowledge area, familiarize yourself with these sections and figures in the *Handbook*.

- Body Current Limit
- Step Voltage
- Touch Voltage
- Substation Grounding Resistance
- Schwarz's Formula
- Schwarz's Formula: Figure A
- Schwarz's Formula: Figure B

The following equations and figures are relevant for knowledge area 1.B.7, Grounding.

Permissible Body Current Limit

Handbook: Body Current Limit

$$I_B = \frac{0.116}{\sqrt{t_s}} \quad \text{[for persons weighing} \approx 50 \text{ kg (110 lb)]}$$

$$I_B = \frac{0.157}{\sqrt{t_s}} \quad \text{[for persons weighing} \approx 70 \text{ kg (155 lb)]}$$

The body current limit is the current a body can withstand and 99.5% of people will survive.

Maximum Allowable Step Voltage

Handbook: Step Voltage

$$E_{\text{step},50} = \left(1000 + 6\,C_s(h_s,\ K)\rho_s\right)\left(\frac{0.116}{\sqrt{t_s}}\right)$$

$$E_{\text{step},70} = \left(1000 + 6\,C_s(h_s,\ K)\rho_s\right)\left(\frac{0.157}{\sqrt{t_s}}\right)$$

$$K = \frac{\rho - \rho_s}{\rho + \rho_s}$$

The step voltage is the difference in surface potential a person could experience while bridging 1 m with the feet without contacting any other ground object.

$E_{\text{step},50}$ represents the step voltage for persons weighing 50 kg (110 lb), and $E_{\text{step},70}$ represents the step voltage for persons weighing 70 kg (155 lb).

The formula for K is based on the surface protective layer, ρ_s, and the soil resistivity beneath.

If there is no protective surface layer, then $C_s = 1$. Otherwise, C_s is estimated from the equation shown.

$$C_s = 1 - \frac{0.09\left(1 - \dfrac{\rho}{\rho_s}\right)}{2h_s + 0.09}$$

A figure is also shown in *Handbook* section Step Voltage that can be used to determine C_s.

Maximum Allowable Touch Voltage

Handbook: Touch Voltage

$$E_{\text{touch},50} = \left(1000 + 1.5\,C_s(h_s,\ K)\rho_s\right)\left(\frac{0.116}{\sqrt{t_s}}\right)$$

$$E_{\text{touch},70} = \left(1000 + 1.5\,C_s(h_s,\ K)\rho_s\right)\left(\frac{0.157}{\sqrt{t_s}}\right)$$

$$K = \frac{\rho - \rho_s}{\rho + \rho_s}$$

The touch voltage is the potential difference between the ground potential rise (GPR) and the point of contact with a person.

The GPR is the maximum electrical potential a substation grid may attain relative to a grounding point assumed to be at the potential of remote earth.

$E_{\text{touch},50}$ represents the touch voltage for persons weighing 50 kg (110 lb), and $E_{\text{touch},70}$ represents the touch voltage for persons weighing 70 kg (155 lb).

Minimum Value of Substation Grounding Resistance in Uniform Soil for Grounding Grid Depth < 0.25 m

Handbook: Substation Grounding Resistance

$$R_g = \frac{\rho}{4}\sqrt{\frac{\pi}{A}}$$

Upper Limit of Substation Grounding Resistance in Uniform Soil for Grounding Grid Depth < 0.25 m

Handbook: Substation Grounding Resistance

$$R_g = \frac{\rho}{4}\sqrt{\frac{\pi}{A}} + \frac{\rho}{L}$$

Substation Grounding Grid Resistance for Grid Depths Between 0.25 and 2.5 m

Handbook: Substation Grounding Resistance

$$R_g = \rho \left[\frac{1}{L} + \frac{1}{\sqrt{20A}} \left(1 + \frac{1}{1 + h\sqrt{\frac{20}{A}}} \right) \right]$$

Total Resistance of a System Consisting of a Combination of Horizontal Grid and Vertical Rods

Handbook: Schwarz's Formula

$$R_g = \frac{R_1 R_2 - R_{12}^2}{R_1 + R_2 - 2R_{12}}$$

$$R_1 = \left(\frac{\rho_1}{\pi L_1} \right) \left[\ln \frac{2L_1}{h} + K_1 \left(\frac{L_1}{\sqrt{A}} \right) - K_2 \right]$$

$$R_2 = \left(\frac{\rho_a}{2n\pi L_2} \right) \left[\ln \frac{8L_2}{d_2} - 1 + 2K_1 \left(\frac{L_2}{\sqrt{A}} \right) \sqrt{n} - 1 \right]$$

$$R_{12} = \left(\frac{\rho_a}{\pi L_1} \right) \left[\ln \frac{2L_1}{L_2} + K_1 \left(\frac{L_1}{\sqrt{A}} \right) - K_2 + 1 \right]$$

K_1 and K_2 are the constants related to the geometry of the system. See *Handbook* figures Schwarz's Formula: Figure A and Schwarz's Formula: Figure B to determine the value of these constants.

Resistance of a Single Ground Rod

Handbook: Schwarz's Formula

$$R_{\text{rod}} = \left(\frac{\rho}{2\pi L} \right) \left(\ln \frac{8L}{d} - 1 \right)$$

Effective Resistance of a Vertical Rod Encased in Concrete

Handbook: Schwarz's Formula

$$R_{\text{CE-rod}} = \left(\frac{1}{2\pi L} \right) \left(\rho_c \ln \frac{D}{d} + \rho \ln \frac{8L}{D} - 1 \right)$$

9. DEFINITIONS

availability: A measure of the percentage of time a machine is operable.

basic lightning impulse insulation level (BIL): The insulation level specified in terms of the crest value of a standard lightning impulse.

body current limit: The current a body can withstand and 99.5% of people will survive.

cavity ratio: A number representative of the geometry of the part of the room between the working plane and the plane of the luminaires.

coefficient of utilization (CU): A measure of the efficiency of a luminaire in transferring luminous energy to a working plane in a particular area.

coincidence factor: The peak of a system divided by the sum of peak loads of its individual components. It tells how likely it is the individual components are peaking at the same time.

demand factor: The ratio of the maximum demand of the demand of the system to the total connected load. It is always less than one (DF < 1).

depreciation: A reduction in the value of an asset with the passage of time.

diversity factor: The ratio of the individual maximum demands of the various subdivisions of a system, or part of a system, to the maximum demand of the whole system. It is usually more than one (DF > 1).

doubling time: The time it takes for energy usage to double.

future worth: The worth of a current asset at a future date based on an assumed rate of growth.

gradient cash flow: Cash flows that increase or decrease by a constant amount. The amount of increase (or decrease) is called the gradient.

ground potential rise (GPR): The maximum electrical potential a substation grid may attain relative to a grounding point assumed to be at the potential of remote earth.

light loss factor: A multiplier that is used to predict future performance (maintained illuminance) based on the initial properties of a lighting system.

lightning impulse insulation level: The insulation level specified for the crest value of a lightning impulse withstand voltage.

mean time between failures (MTBF): The arithmetic mean (average) time between predicted failures of a system for a repairable system.

mean time to failure (MTTF): The expected time to failure for a non-repairable system.

mean time to repair (MTTR): The average time to repair a piece of equipment.

present worth: The amount of cash today that is equivalent in value to a payment, or number of payments, to be received in the future.

protection quality index: A measure of how well a surge arrester can protect a system.

protective margin: The effectiveness of a surge arrester for a particular system.

reliability: The probability that an item will continue to operate satisfactorily up to a time, t.

reseal voltage: The voltage at which the arrester regains a high impedance.

standard lightning impulse: A full impulse with a front time of 1.2 μs and a half-value time of 50 μs. This is normally called a 1.2/50 impulse.

step voltage: The difference in surface potential a person could experience while bridging 1 m with the feet without contacting any other ground object.

stroke: Any one of a series of repeated discharges that comprise a single discharge, also known as a *lightning flash* or *strike*.

surge: A transient with a short duration and a high magnitude. Surges are caused by lightning strikes or switching operations.

surge arrester: An overvoltage protection device that temporarily connects a circuit to ground during an overvoltage until the energy is dissipated.

surge impedance loading: The power load in which the total reactive power of the line becomes zero.

touch voltage: The potential difference between the ground potential rise (GPR) and the point of contact with a person.

voltage protection level: The maximum voltage allowed by a surge arrester.

10. NOMENCLATURE

a	per unit growth rate	–
A	annual value	$
A	area	m^2 (ft^2)
A	availability	–
B	benefit	$
BV	book value	$
C	coefficient factor	–
C	cost	$
C	layer derating factor	–
CCR	ceiling cavity ratio	–
CR	cavity ratio	–
CU	coefficient of utilization	–
d	deflated interest rate period	time
d	diameter of ground rod	m
d	distance	m (ft)
D	coincidence demand of group	–
D	demand	–
D	depreciation	$
D	diameter of concrete shell	m
DF	demand factor	–
e	decimal inflation rate	$
E	energy	J
E	illuminance	lx (fc)
E	voltage	V
f	inflation rate	–
F	factor	–
F	future worth	$
FCR	floor cavity ratio	–
FLD	load factor	–
FLS	loss factor	–
G	gradient cash flow	$
h	thickness of material	m
h, H	height, thickness	m
i	interest rate	%
i'	effective interest rate	–
I	current	A
I	luminous intensity	cd
IL	illumination level	–
K	geometry constant	–
L	length	m
L	luminance	foot-lamberts
LC	luminance coefficient	–
LD	load diversity	VA
LDD	luminaire dirt depreciation	–
LLD	lumen lamp depreciation	–
LLF	light loss factor	–
m	number of time intervals	–
MH	mounting height	ft
MMI	minimum maintained illumination	fc
MTBF	mean time before failure	hr
MTTF	mean time to failure	hr
MTTR	mean time to repair	hr
n	number of years	–
N	number	–
ND	mounting height constant	–
P	power	W
P	present worth	$
P, p	probability	–
PQI	protection quality index	–
PM	protection margin	–
Pf	plant factor	–
r	rate	%
R	reliability	–
RCR	room cavity ratio	–
RRC	reflected radiation coefficient	–
S	salvage value	$
SIL	surge impedance loading	Ω

t	time	s
T	period	s
V	voltage	V
W	width	m
Z	impedance	Ω

Symbols

ϵ	permittivity	F/m
η	efficiency	%
θ	angle of incidence	rad
λ	failure rate	failures/hr
μ	permeability	H/m
ρ	resistivity	$\Omega \cdot m$
Φ	probability	–

Subscripts

0	characteristic, current, free space, initial, or vacuum
al	annual load
avg	average
B	body
c	coincidence compounding, or concrete
C	characteristic
cc	ceiling cavity
CE	concrete, effective
d	doubling
D	density
div	diversity
fc	floor cavity
g	ground or group
G	ground flash
in	input
j	number of years
L	line or loss
max	maximum
n	year
out	output
p	protection
r	relative or reseal
rc	room cavity
s	seconds
s	soil, surface, or surge
u	utilization
w	wall or withstand

3 Codes and Standards

Exam specification 1.C, Codes and Standards, makes up between 13% and 19% of the PE Electrical Power exam (between 10 and 15 questions out of 80).

The organization of this chapter follows the order of knowledge areas given by the NCEES for this exam specification. Each knowledge area is covered in the following numbered sections.

Content in blue refers to the *NCEES Handbook*.

Content in red is additional essential information.

While the *NCEES Handbook* does not specifically include information on the knowledge areas within exam specification 1.C, Codes and Standards, there are several resources available for you to consult.

During the exam, you will be provided copies of the following codes.

- *National Electrical Code* (NFPA 70, NEC-2017)

- *National Electrical Safety Code* (ANSI C2, NESC-2017)

- *Standard for Electrical Safety in the Workplace: Shock and Burns* (NFPA 70E-2018)

- *Recommended Practice for the Classification of Flammable Liquids, Gases, or Vapors and of Hazardous (Classified) Locations for Electrical Installations in Chemical Process Areas* (NFPA 497-2017)

- *Recommended Practice for the Classification of Combustible Dusts and of Hazardous (Classified) Locations for Electrical Installations in Chemical Process Areas* (NFPA 499-2017)

- *Code for the Manufacture and Storage of Aerosol Products* (NFPA 30B-2015)

1. NATIONAL ELECTRICAL CODE (NFPA 70, NEC-2017)

The *National Electrical Code* (NEC) will be supplied during testing. However, it is imperative that you understand its contents and usage well before the exam. To that end, we recommend the purchase or download of the *NEC Handbook*. The *NEC Handbook* has information on updates, explains portions of the NEC, provides calculational examples, and provides pictures, drawings, and figures that are invaluable to understanding and using the NEC.

Knowledge Area Overview

Key concepts: These key concepts are important for answering exam questions in knowledge area 1.C.1, National Electrical Code (NFPA 70, NEC-2017).

- breaker size and conductor derating calculations

- familiarity with important NEC tables

- finding information quickly in the NEC

- voltage drop calculations

- units of circular mils

PE Power Reference Manual **(EPRM):** Study these sections in EPRM that either relate directly to this knowledge area or provide background information.

- Section 7.26: Magnetic Field Strength

- Section 16.3: Resistance

- Section 26.4: Overcurrent Protection

- Section 39.1: History and Overview

- Section 41.12: Insulation and Ground Testing

- Chapter 44: National Electrical Code

The NEC consists of an introduction followed by nine chapters, which are further subdivided into articles. The outline of the NEC is shown here. Chapter numbers are given in parentheses. Chapter 1 through Chap. 4 generally apply to all situations except as modified by Chap. 5 through Chap. 7.

- Introduction

- (1) General

- (2) Wiring and Protection

- (3) Wiring Methods and Materials

- (4) Equipment for General Use

- (5) Special Occupancies

- (6) Special Equipment

- (7) Special Conditions

- (8) Communications Systems

- (9) Tables

- Informative Annexes

NCEES Handbook: To prepare for this knowledge area, familiarize yourself with this section in the Handbook.

- Voltage Drop

The following equations, figure and NEC articles and tables are relevant for knowledge area 1.C.1, National Electrical Code (NFPA 70, NEC-2017).

NEC Chapter 1: General

EPRM: Sec. 44.3

- Article 100: Definitions

  EPRM Fig. 44.2 shows a graphic depiction of some of the definitions in Article 100.

  Article 100 is always a great place to start with any code problem. Not all definitions will be here, but when the code has a specific meaning, check here first.

Figure 44.2 Typical Distribution System

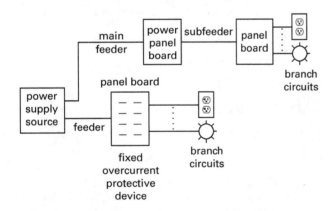

- Article 110: Requirements for Electrical Installations

  Article 110 contains requirements that apply across most applications. Skim the article and read the commonly used requirements. NEC Table 110.28, Enclosure Selection, gives a good chart of which

enclosures to use for electrical components depending on the conditions.

NEC Chapter 2: Wiring and Protection

EPRM: Sec. 44.4–Sec. 44.10

Classic questions on the PE exam that are relevant to NEC Chap. 2 include the following.

- voltage drop

- sizing a conductor

- sizing an overcurrent protective device

- derating for multiple conductors and ambient temperatures

- calculating values for motor circuits

NEC Chap. 2 contains several important requirements that may come up on the exam.

- Article 200: Use and Identification of Grounded Conductors

- Article 210: Branch Circuits

  NEC Art. 210 contains a lot of important requirements and tables that may come up during the exam.

  NEC Sec. 210.11, Branch Circuits Required, is a typical starting point for construction requirements for household and other living spaces.

  NEC Sec. 210.12, Arc-Fault Circuit-Interrupter Protection, includes locations where arc-fault circuit-interrupter (AFCI) protection is required.

  NEC Sec. 210.13, Ground-Fault Protection of Equipment, is important to read, especially the commentary note.

  Review NEC Sec. 210.19: Conductors—Minimum Ampacity and Size.

  Review NEC Sec. 210.20, Overcurrent Protection, for recommended voltage drop values from branch circuits.

- Article 215: Feeders

  NEC Sec. 215.2(A)(1)(b), Informational Note 2 contains the maximum recommended voltage drop: 3% for feeders and a total of 5% for feeders and branch circuits.

  Consult NEC Sec. 215.2 and Sec. 215.3 for load capacity for feeder conductors.

- Article 220: Branch-Circuit, Feeder, and Service Load Calculations

  Article 220 is the basis for many of the computational problems on the exam and is important to study. The tables in Article 220 may be needed in calculations.

- Article 225: Outside Branch Circuits and Feeders

  NEC Art. 225 is not likely to be on the exam, so skim this section.

- Article 230: Services

  NEC Art. 230 is worth reading, though not as likely to be on the exam as other sections.

- Article 240: Overcurrent Protection

  NEC Art. 240 contains overcurrent protection requirements, which are fairly likely to be on the exam.

  NEC Sec. 240.6, Standard Ampere Ratings, contains standard ratings for fuses and inverse time circuit breakers. The standard ratings for fuses and fixed-trip circuit breakers are given in NEC Sec. 240.6(A) and NEC Table 240.6(A).

- Article 250: Grounding and Bonding

  NEC Art. 250 contains grounding and bonding requirements, which are applicable to every electrical installation and are frequently misunderstood and misapplied.

  Review NEC Table 250.122, Equipment Grounding.

- Article 280: Surge Arresters, Over 1000 Volts

  NEC Art. 280 is a short section and should be read.

- Article 285: Surge-Protective Devices (SPDs), 1000 Volts or Less

  NEC Art. 285 is a short section and should be read.

NEC Chapter 3: Wiring Methods and Materials

EPRM: Sec. 44.11

NEC Chap. 3 is very important because it contains ampacity tables.

- Article 300: General Requirements for Wiring Methods and Materials

  Skim NEC Art. 300.

- Article 310: Conductors for General Wiring

  Read NEC Art. 310.

  The 2017 NEC does not have the conductor designations, but the exam might require knowing the selection criteria for conductors in NEC Sec. 310.10.

  Be familiar with NEC Table 310.15(B)(2)(a), Ambient Temperature Correction Factors Based on 30°C (86°F), and NEC Table 310.15(B)(2)(b), Ambient

Temperature Correction Factors Based on 40°C (104°F).

Ampacity of conductors must be adjusted according to NEC Table 310.15(B)(3)(a) when more than three conductors are housed in a raceway.

Allowable conductor ampacities are found in NEC Table 310.15(B)(16). The ampacities apply to a maximum of three conductors in a raceway.

The rest of NEC Chap. 3 contains code requirements that an electrician would need to know and are unlikely to come up on the exam.

NEC Chapter 4: Equipment for Use

EPRM: Sec. 44.12

NEC Art. 410 and 430 are the most important articles in Chap. 4. The rest of the chapter can be skimmed.

- Article 410: Luminaires, Lampholders, and Lamps

  Read NEC Art. 410.

- Article 430: Motors, Motor Circuits, and Controllers

  Read NEC Art. 430.

  Review NEC Table 430.250 because it may be used in calculation problems.

- Article 450: Transformers and Transformer Vaults (Including Secondary Ties)

  Review NEC Table 450.3(A) for the maximum rating for an overcurrent device for transformers over 1000 V.

  Review NEC Table 450.3(B) for the maximum overcurrent protection for transformers 1000 V and less.

- Article 460: Capacitors

  This article covers the installation of capacitors on electrical circuits.

- Article 480: Storage Batteries

  The provisions of this article apply to all stationary installations of storage batteries.

- Article 490: Equipment Over 1000 Volts, Nominal

NEC Chapter 5: Special Occupancies

EPRM: Sec. 44.13

Chapter 5 may be relevant if there are problems based on NFPA 497, 499, or 30B. Students should read the articles that relate to NFPA 497, 499, and 30B.

NEC Chapters 6–8

EPRM: Sec. 44.14–Sec. 44.16

Chapters 6–8 are not likely to be on the exam. Students should be familiar with these chapters, but in-depth study is not necessary.

NEC Chapter 9: Tables

EPRM: Sec. 44.17

Chapter 9 contains several tables that are essential to solving many of the problems on the exam. Students should know which tables are in this chapter and how to use them.

The following are some of the more commonly used tables.

- NEC Table 5A, "Compact Copper and Aluminum Building Wire Nominal Dimensions and Areas"

- NEC Table 8, "Conductor Properties"

- NEC Table 9, "Alternating-Current Resistance and Reactance for 600-Volt Cables, 3-Phase, 60 Hz, 75°C (167°F)—Three Single Conductors in Conduit"

NEC Annexes

EPRM: Sec. 44.18

There are ten NEC annexes.

- Informative Annex A: Product Safety Standards

- Informative Annex B: Application Information for Ampacity Calculation

- Informative Annex C: Conduit and Tubing Fill Tables for Conductors and Fixture Wires of the Same Size

- Informative Annex D: Examples

  Review Informative Annex D, Example D8, Motor Circuit Conductors, Overload Protection, and Short-Circuit and Ground-Fault Protection.

  Load calculations could be asked for on the PE exam. Informative Annex D will help with these calculations.

- Informative Annex E: Types of Construction

- Informative Annex F: Availability and Reliability for Critical Operations Power Systems; and Development and Implementation of Functional Performance Tests (FPTs) for Critical Operations Power Systems

- Informative Annex G: Supervisory Control and Data Acquisition (SCADA)

- Informative Annex H: Administration and Enforcement

- Informative Annex I: Recommended Tightening Torque Tables from UL Standard 486A-B

- Informative Annex J: ADA Standards for Accessible Design

Two-Wire Circuit Voltage Drop

Handbook: **Voltage Drop**

$$\text{VD} = \frac{2LRI}{K(1000)}$$

EPRM: Sec. 44.7

$$V_{\text{drop}} = \frac{2lIR}{1000 \text{ ft}} = IR_l 2l \qquad \textit{44.4}$$

Voltage drop calculations are important and necessary to know for the PE exam. The *Handbook* and EPRM equations are equivalent for single-phase AC or DC circuits, where $K = 1.0$.

The numeral two (2) is included in each equation because the voltage drop occurs both in the outgoing and incoming wires.

The power factor is considered to be equal to one (1) in this formula; it is purely resistive.

The 1000 ft is used because the NEC tables in Chap. 9 use "per 1000 ft" values. See NEC Sec. 215.2(A)(1)(b), Informational Note 2, for the maximum recommended voltage drop.

The resistance value, R, is obtained from NEC Chap. 9, Table 9. The unit for R is Ω/1000 ft or km.

There are additional explanation and examples at the end of NEC Sec. 215.2(A)(1) Informational Notes 1 and 2 and following NEC Chap. 9, Table 9.

Area of Circular Mil

EPRM: Sec. 44.3

$$A_{\text{cmil}} = \left(\frac{d_{\text{in}}}{0.001}\right)^2 \qquad \textit{44.1}$$

This is the mathematical definition of circular mil, an area of one mil which is equal to an area of wire with a diameter of one one-thousandth of an inch, or 0.001 inches.

Conductor sizes are referred to in American Wire Gauge (AWG) or in circular mil size in the NEC tables.

Conductor Ampacity

EPRM: Sec. 44.11

The selection of conductor ampacity involves detailed calculations and consideration of many limitations and adjustments. In addition to the information in EPRM Sec. 44.11, there exists a large number of NEC articles and tables in multiple NEC chapters that addresses the selection of conductor ampacity. A brief list of frequently referenced NEC articles and annexes includes

- NEC Art. 110
- NEC Art. 210
- NEC Art. 220
- NEC Art. 310
- NEC Art. 410
- NEC Art. 430
- Informative Annex B

The process steps for selection of conductor ampacity are given in the EPRM Sec. 44.11 and is worthy of detailed study.

2. NATIONAL ELECTRICAL SAFETY CODE (ANSI C2, NESC-2017)

The exam may present a problem on any of the several tables of data in the *National Electrical Safety Code* (NESC). Students are advised to purchase the NESC.

Knowledge Area Overview

Key concepts: These key concepts are important for answering exam questions in knowledge area 1.C.2, National Electrical Safety Code (ANSI C2, NESC-2017).

- familiarity with important NESC tables
- finding information quickly in the NESC
- typical calculations such as maximum height and required vertical clearance for electrical equipment

PE Power Reference Manual **(EPRM):** Study these sections in EPRM that either relate directly to this knowledge area or provide background information.

- Chapter 45: National Electrical Safety Code

The NESC consists of general sections followed by four parts (indicated by single digits), all of which are further subdivided into section numbers (indicated by two digits) and then rules (indicated by three digits). An outline of the top-level structure of the NESC is shown here.

- General Sections
- Part 1: Safety Rules for the Installation and Maintenance of Electric Supply Stations and Equipment
- Part 2: Safety Rules for the Installation and Maintenance of Overhead Electric Supply and Communication Lines
- Part 3: Safety Rules for the Installation and Maintenance of Underground Electric Supply and Communication Lines
- Part 4: Work Rules for the Operation of Electric Supply and Communication Lines and Equipment
- Appendices

NCEES Handbook: The *Handbook* does not include any sections for this knowledge area.

The following NESC sections and rules are relevant for knowledge area 1.C.2, National Electrical Safety Code (ANSI C2, NESC-2017).

NESC Areas of Concern

The NESC is a performance code, not a design specification. It specifies what is to be performed, not how. The NESC is often adopted by state authority with jurisdiction over utilities, and generally applies to utilities.

NESC areas of concern are as follows.

- electric supply stations
- overhead lines
- underground lines
- work rules

NESC Areas of Applicability

The NESC covers the part of the electric system from and (including the generation facility) up to the service entrance where the NEC takes jurisdiction. The area of coverage for the safety code is shown in EPRM Fig. 45.2.

The NESC no longer covers electric fences, radio installations, or utilization equipment as these are now in the NEC, nor does it cover mines or ships. (See IEEE Std. 43 for Electrical Installations Aboard Ships.)

NESC: General Sections

EPRM: Sec. 45.2–Sec. 45.5

The general sections are as follows.

- (01) Introduction
- (02) Definitions

- (03) References
- (09) Grounding

  Grounding conductors should be copper-covered steel (Rule 093A).

  Ground resistance should be 25 Ω or less (Rule 096).

NESC Part 1: Safety Rules for the Installation and Maintenance of Electric Supply Stations and Equipment

EPRM: Sec. 45.6

Part 1 covers the practical safeguarding of persons during the installation, operation, or maintenance of electric supply stations and their associated equipment. Students should be familiar with many requirements in Part 1.

The sections in Part 1 are as follows.

- (10) Purpose and Scope of Rules
- (11) Protective Arrangements in Electric Supply Stations

  Section 11 covers rules on enclosures, which students should be familiar with.

- (12) Installation and Maintenance of Equipment
- (13) Rotating Equipment
- (14) Storage Batteries
- (15) Transformers and Regulators
- (16) Conductors
- (17) Circuit Breakers, Reclosers, Switches, and Fuses
- (18) Switchgear and Metal-Enclosed Bus
- (19) Surge Arresters

  Surge arresters protect against indirect lightning strikes. Section 19 covers rules on the installation and location of surge arresters (Rules 190, 191, and 193).

NESC Part 2: Safety Rules for the Installation and Maintenance of Overhead Electric Supply and Communication Lines

EPRM: Sec. 45.7

Part 2 covers the practical safeguarding of persons during the installation, operation, or maintenance of overhead supply and communication lines and their associated equipment. The exam may have questions on grounding, clearances, conductor spacing, and strength.

The sections in Part 2 are as follows.

- (20) Purpose, Scope, and Application of Rules
- (21) General Requirements

- (22) Relations Between Various Classes of Lines and Equipment
- (23) Clearance

  Section 23 is about required clearances and is likely to be on the exam.

- (24) Grades of Construction

  Section 24 is about choosing conductors that are strong enough for safety under the conditions.

- (25) Loadings for Grades B and C

  Section 25 discusses expected loading of overhead lines in regions where wind and ice accumulation on overhead lines may be expected.

- (26) Strength Requirements
- (27) Line Insulation

NESC Part 3: Safety Rules for the Installation and Maintenance of Underground Electric Supply and Communication Lines

EPRM: Sec. 45.8

Part 3 covers the practical safeguarding of persons during the installation, operation, or maintenance of underground or buried supply and communications cables and associated equipment. Part 3 covers information that is largely specific to underground supply.

The sections in Part 3 are as follows.

- (30) Purpose, Scope, and Application of Rules
- (31) General Requirements Applying to Underground Lines
- (32) Underground Conduit Systems

  Rule 320 deals with the location of underground conduit systems, where they can be routed, and how far they must be separated from other underground installations. The NESC does not specify conduit burial depths, but instead uses the phrase "accepted good practice," which generally means that the requirements of the NEC are applicable.

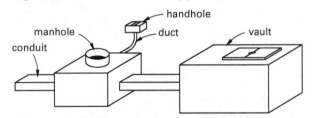

conduit system (combination of conduits, manholes, handholes, and/or vaults joined to form an integrated whole)

Rule 320B.2 covers underground supply (power) and communication conduit separation requirements. The data in EPRM Table 45.2 comes from this NESC rule.

Review NESC Table 341-1, "Clearance Between Supply and Communications Facilities in Joint-Use Manholes and Vaults." The data in EPRM Table 45.3 comes from this NESC table.

- (33) Supply Cable
- (34) Cable in Underground Structures
- (35) Direct-Buried Cable and Cable in Duct Not Part of a Conduit System

The NESC contains burial depth requirements for direct-buried supply cables, but not communication cables. These depth requirements are covered in Rule 352 and shown in EPRM Fig. 45.6(a). EPRM Fig. 45.6 comes from NESC Table 352-1, Supply Cable, Conductor, or Duct Burial Depth. The depth varies with the voltage and is measured surface to surface, not center to center.

- (36) Risers
- (37) Supply Cable Terminations
- (38) Equipment
- (39) Installation in Tunnels

NESC Part 4: Work Rules for the Operation of Electric Supply and Communications Lines and Equipment

EPRM: Sec. 45.9

Part 4 provides practical work rules on the means of safeguarding employees and the public from injury. It is not the intent of the code to impose unreasonable rules; however, all reasonable steps should be taken.

The sections in Part 4 are as follows.

- (40) Purpose and Scope
- (41) Supply and Communications Systems—Rules for Employers
- (42) General Rules for Employees
- (43) Additional Rules for Communications Employees
- (44) Additional Rules for Supply Employees

NESC Appendices

EPRM: Sec. 45.4

There are two appendices that cover clearances. There is no reason to study the appendices.

- (A) Uniform System of Clearances
- (B) Uniform Clearance Calculations for Conductors Under Ice and Wind Conditions
- (C) Example Applications for Rule 250C Tables 250-2 and 250-3
- (D) Determining Maximum Anticipated Per-Unit Over-voltage Factor (T) at the Worksite

3. STANDARD FOR ELECTRICAL SAFETY IN THE WORKPLACE: SHOCK AND BURNS (NFPA 70E-2018)

The National Fire Protection Association *Standard for Electrical Safety in the Workplace*, (NFPA 70E), gives general guidelines and some specific requirements for electrical safety programs for facilities. The *PE Power Reference Manual* (EPRM) focuses on the shock and burn aspects of these requirements as well as the personal protective equipment (PPE) required.

Knowledge Area Overview

Key concepts: These key concepts are important for answering exam questions in knowledge area 1.C.3, Standard for Electrical Safety in the Workplace: Shock and Burns (NFPA 70E-2018).

- effects of shock and burns
- finding information quickly in NFPA 70E-2018

PE Power Reference Manual **(EPRM):** Study these sections in EPRM that either relate directly to this knowledge area or provide background information.

- Section 31.15: Electrical Safety in the Workplace
- Section 31.16: Arc-Flash and Shock Hazard Levels
- Section 43.6: Shock and Burns

The outline of the NFPA 70E is shown here.

- Introduction
- Chapter 1: Safety-Related Work Practices
- Chapter 2: Safety-Related Maintenance Requirements
- Chapter 3: Safety Requirements for Special Equipment
- Informative Annexes

NCEES Handbook: The *Handbook* does not include any sections for this knowledge area.

The following tables, figures, and standards are relevant for knowledge area 1.C.3, Standard for Electrical Safety in the Workplace: Shock and Burns (NFPA 70E-2018).

NFPA 70E Chapter 1: Safety-Related Work Practices

EPRM: Sec. 31.15

Chapter 1 contains basic principles for safety programs.

- requirements for work performed on energized electrical equipment

- lockout/tagout procedures

- establishing an electrical safety program for the installation

- work permits

- performing shock and arc-flash analysis on electrical installations to identify hazards and risks

- using proper PPE

- installing necessary protective shields or barriers

- training maintenance and operations workers, as well as those exposed to such work

NFPA 70E Chapter 2: Safety-Related Maintenance Requirements

EPRM: Sec. 31.15

Chapter 2 has many requirements for maintenance. The requirements identify only maintenance directly associated with employee safety.

NFPA 70E Chapter 3: Safety Requirements for Special Equipment

EPRM: Sec. 31.15

Training guidance and requirements, arc-flash assessment and PPE, all of which could be tested, are covered in this chapter. Review the flow charts provided as aids to determining requirements.

NFPA 70E Informative Annexes

EPRM: Sec. 31.15

The annexes are for information only and contain no requirements. Nevertheless, they are excellent sources of information for understanding the requirements themselves.

Approach Boundary Requirements

EPRM: Sec. 31.16

The protection in NFPA 70E is based on the distance from the hazard (represented by the approach boundaries), PPE required, and the qualifications of the individuals to be in each area.

There are two types of boundaries, shock boundaries—which are restricted or limited—and arc-flash boundaries.

Figure 31.21 Approach Boundaries

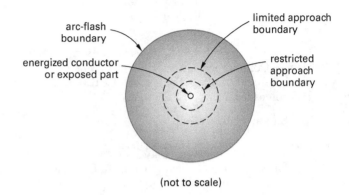

(not to scale)

Table 31.2 Approach Boundary Limits to Energized Components

nominal system voltage range, phase to phase	limited approach boundary		restricted approach boundary
	exposed movable conductor	exposed fixed circuit part	
AC < 50 V	not specified	not specified	not specified
50 V ≤ AC ≤ 150 V	3.0 m (10 ft 0 in)	1.0 m (3 ft 6 in)	avoid contact
DC < 100 V	not specified	not specified	not specified
100 V ≤ DC ≤ 300 V	3.0 m (10 ft 0 in)	1.0 m (3 ft 6 in)	avoid contact

Adapted from NFPA 70E®-2018, *Standard for Electrical Safety in the Workplace*, © 2018, National Fire Protection Association.

See NFPA 70E Table 130.4(D)(a) for a complete table on AC systems.

See NFPA 70E Table 130.4(D)(b) for a complete table on DC systems.

Shock Impacts and Current Levels

EPRM: Sec. 43.6

NFPA 70E defines a shock hazard as "a dangerous condition associated with the possible release of energy."

The shock impacts not only vary by person, but also by type of current (e.g., static discharge, DC, 60 Hz AC, 400 Hz AC or higher). The let-go values also vary by the standard used.

Protective equipment is based on the amount of energy released but is given in calories and not in amperes, as in Fig. 43.2.

Figure 43.2 *Human Physiological Reactions*

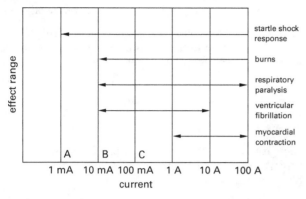

A = perception
B = let go
C = death

4. HAZARDOUS AREA CLASSIFICATION (NFPA 497-2017, 499-2017, 30B-2015)

National Electrical Code (NEC) Art. 500 contains the requirements to ensure that electrical equipment within a hazardous area will not ignite flammable or combustible material; however, the actual classification of the area is left to other standards, which are reviewed in this knowledge area.

Knowledge Area Overview

Key concepts: These key concepts are important for answering exam questions in knowledge area 1.C.4, Hazardous Area Classification (NFPA 497-2017, 499-2017, 30B-2015).

- finding information quickly in National Fire Protection Association codes NFPA 497-2017, 499-2017, and 30B-2015

- hazardous area classifications

PE Power Reference Manual **(EPRM):** Study these sections in EPRM that either relate directly to this knowledge area or provide background information.

- Section 44.19: Hazardous Area Classification

- Section 44.20: Hazardous Area Classification: Aerosol Products (NFPA 30B)

- Section 44.21: Hazardous Area Classification: Liquids, Gases, and Vapors (NFPA 497)

- Section 44.22: Hazardous Area Classification: Combustible Dust (NFPA 499)

The classification of hazardous areas is covered in the following codes.

- NFPA 30B: *Code for the Manufacture and Storage of Aerosol Products*, 2015. National Fire Protection Association.

- NFPA 497: *Recommended Practice for the Classification of Flammable Liquids, Gases, or Vapors and of Hazardous (Classified) Locations for Electrical Installations in Chemical Process Areas*, 2017. National Fire Protection Association.

- NFPA 499: *Recommended Practice for the Classification of Combustible Dusts and of Hazardous (Classified) Locations for Electrical Installations in Chemical Process Areas*, 2017. National Fire Protection Association.

NCEES Handbook: The *Handbook* does not include any sections for this knowledge area.

The following table and standards are relevant for knowledge area 1.C.4, Hazardous Area Classification (NFPA 497-2017, 499-2017, 30B-2015).

Liquids, Gases, and Vapors (NFPA 497-2017)

EPRM: Sec. 44.21

NFPA 497 has a table of the properties of many (though not all) flammable gases or vapors, flammable liquids, and combustible liquids; this table is used to determine the hazard classification per the NEC.

Read Chap. 1 and Chaps. 3–5 (there isn't much to read, though; it is mostly tables and figures, whose titles and coverage you should be familiar with).

NEC Art. 500 and 505 are related to NFPA 497.

Combustible Dust (NFPA 499-2017)

EPRM: Sec. 44.22

NFPA 499 has a table of the properties of many (though not all) flammable dusts; this table is used to determine the hazard classification per the NEC.

Read Chap. 1 and Chaps. 3–6 (there isn't much to read; it is mostly tables and figures).

NEC Art. 500 and 506 are related to NFPA 499.

Aerosol Products (NFPA 30B-2015)

EPRM: Sec. 44.20

NFPA 30B covers the manufacture, storage, and display of aerosol products. It applies to specific types of aerosol containers. The types of aerosols and containers should be understood as they will be relevant to classification questions on the exam.

See NFPA 30B Sec. 1.1.1–1.1.4 for details.

Classes, Divisions, and Zones

EPRM: Sec. 44.19

Each standard breaks the requirements into a *class* based on the properties of the flammable gas or vapor (Class I), *combustible dust* (Class II), or fibers/flyings (Class III) that might be present. Each class is further subdivided into a *division* depending, generally, on whether the hazard is present during normal operations (Division 1) or present during abnormal conditions (Division 2). NEC Art. 500 incorporates the standards from NFPA 497, NFPA 499, and NFPA 30B and covers Classes I, II, and III with Divisions 1 and 2. Classes can also be divided into divisions or zones. The equivalency is shown in EPRM Table 44.2.

Table 44.2 Group Equivalency Between Division and Zone Classifications

division system groups	zone system groups
IIC	A and B
IIB	C
IIA	D
I	D

5. DEFINITIONS

arc fault circuit interrupter: A device intended to protect from the effects of an arc fault by recognizing the unique characteristics of an arc and de-energizing the circuit when said characteristics are detected.

arc-flash boundary: The approach limit at a distance from an arc source within which a person could receive a second-degree burn if an electrical arc were to occur.

branch circuit: The circuit conductor between the final overcurrent device protecting the circuit and the outlet(s). Branch circuits are divided into four categories: appliance, general purpose, individual and multiwire.

circular mil: An area of wire with a diameter of one one-thousandth, or 0.001, inches.

ground fault circuit interrupter: A GFCI is a device that de-energizes a circuit within an established period of time when the current to ground exceeds a predetermined level, which is less than that required to operate the circuit's overcurrent protective device.

intrinsically safe: A circuit that does not develop sufficient electrical energy (millijoules) in an arc or a spark to cause ignition, or sufficient thermal energy resulting from an overload condition to cause the temperature of the installed circuit to exceed the ignition temperature of a specified gas or vapor under normal or abnormal operating conditions. [Source: *NFPA 70 Handbook*]

let-go level: Current level where humans lose control of their muscles. That is, the muscles contract and a person is unable to let go until the current is removed.

limited approach boundary: Boundary within which, in general, only qualified persons are allowed. Unqualified persons are allowed if supervision requirements are met.

restricted approach boundary: Requires a qualified person and insulation from the hazard, either on the person or on the hazard.

surge arrester: An overvoltage protection device that temporarily connects a circuit to ground during an overvoltage until the energy is dissipated.

6. NOMENCLATURE

A	area	cmil
d	diameter	in
I	current	A
l	length	ft
K	constant	–
R	resistance	Ω
V	voltage	V

Subscripts

cmil	circular mil
in	inches
l	per unit length

4 Analysis

Exam specification 2.A, Analysis, makes up between 10% and 15% of the PE Electrical Power exam (between 8 and 12 questions out of 80).

The organization of this chapter follows the order of knowledge areas given by the NCEES for this exam specification. Each knowledge area is covered in the following numbered sections.

Content in blue refers to the *NCEES Handbook*.

Content in red is additional essential information.

1. THREE-PHASE CIRCUITS

Three-phase circuits provide greater power in smaller volumes than similar single-phase devices. While more than three phases could have been chosen as a standard, additional phases would mean using more copper, resulting in higher cost. Three-phase power is cost-effective for the amount of power delivered. The frequency chosen for three-phase power in the United States is 60 Hz. This frequency is based on minimizing light flicker and the operation of induction motors.

Knowledge Area Overview

Key concepts: These key concepts are important for answering exam questions in knowledge area 2.A.1, Three-Phase Circuits.

- benefits of three-phase power systems

- voltages and phase angles

- three-phase circuit analysis

- determination of various parameters for three-phase circuits

PE Power Reference Manual (**EPRM**): Study these sections in EPRM that either relate directly to this knowledge area or provide background information.

- Section 17.21: Apparent Power

- Section 17.22: Complex Power and the Power Triangle

- Chapter 24: Three-Phase Electricity and Power

- Section 26.1: Fundamentals

- Section 26.3: Common-Neutral System

- Section 26.7: Fault Analysis: Symmetrical

- Section 26.8: Fault Analysis: Unsymmetrical

- Section 27.5: Three-Phase Transformer Configurations

- Section 28.9: Three-Phase Transmission

- Section 29.2: Power Flow

- Section 29.3: Three-Phase Connections

- Section 29.4: Operator: 120°

- Section 29.6: Per-Unit System

- Section 29.7: Sequence Components

- Section 33.13: Synchronous Machine Equivalent Circuit

- Section 41.14: Measurement Standards and Conventions

NCEES Handbook: To prepare for this knowledge area, familiarize yourself with this section in the *Handbook*.

- 3-Phase Circuits

The following equations and figures are relevant for knowledge area 2.A.1, Three-Phase Circuits.

Delta: Voltage and Current Relationships

Handbook: 3-Phase Circuits

EPRM: Sec. 27.5

$$V_L = V_P$$
$$I_L = \sqrt{3}\, I_P$$

Though differing in subscripts, the *Handbook* equations and EPRM Eq. 27.8 and Eq. 27.9 are equivalent. When determining full power, three-phase problems can be solved as single-phase problems and the answer multiplied by three.

Delta-connected line values and phase angles differ from one another in the *Handbook* and EPRM. For example, $+120°$ and $-240°$ are equivalent angles. EPRM Fig 29.7 shows a summary of the angular relationships.

Figure 29.7 *Balanced Delta Current Relationships*

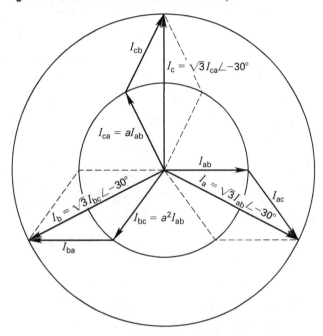

(a) line-to-line currents versus phase currents

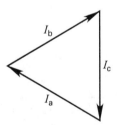

(b) alternative line versus neutral currents

Wye: Voltage and Current Relationships

Handbook: 3-Phase Circuits

EPRM: Sec. 27.5

$$V_L = \sqrt{3}\,V_P = \sqrt{3}\,V_{LN}$$
$$I_L = I_P$$

Though differing in subscripts, the *Handbook* equations and EPRM Eq. 27.10 and Eq. 27.11 are equivalent. When determining full power, three-phase problems can be calculated as single-phase problems and then multiplied by three to obtain a final answer.

Wye line values and phase angles differ from one another in the *Handbook* and EPRM. For example, $+120°$ and $-240°$ are equivalent angles. EPRM Fig. 29.6 shows a summary of the angular relationships.

Figure 29.6 *Balanced Wye Voltage Relationships*

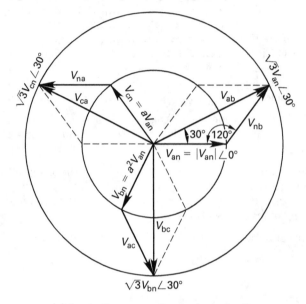

(a) line-to-line versus line-to-neutral voltages

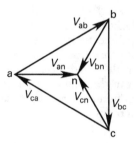

(b) alternative line-to-line versus line-to-neutral voltages

Apparent Power: Rectangular Version

Handbook: 3-Phase Circuits

$$S = P + jQ$$

EPRM: Sec. 29.2

$$\begin{aligned} \mathbf{S} &= \mathbf{V}\mathbf{I}^* \\ &= P + jQ \\ &= VI\cos\theta + jVI\sin\theta \end{aligned}$$

29.1

The apparent power is related to the real and reactive power. Real power, P, uses watts (W); reactive power, Q, uses volt-amperes reactive (VAR); and apparent power, S, uses volt-amperes (VA).

A system must be designed to handle the apparent power, not just the real power. Real power is absorbed by a load. The reactive power is exchanged by the source and the load, and the impact of this exchange must be accounted for in a given design. The defining design criterion is the apparent power, which is the combination of real and reactive power.

The complex conjugate of the current is used to ensure that the result includes the proper angular relationship. The apparent power is the product of the rms voltage and the rms current as shown in EPRM Eq. 41.24.

$$\mathbf{S} = \mathbf{VI}^* = |V||I| \angle (\theta - \phi_I) \qquad \textit{41.24}$$

Normally, the angle associated with the apparent power is not of concern and apparent power is represented as shown in EPRM Eq. 17.57.

$$S = IV \qquad \textit{17.57}$$

The relationship between apparent power and the impedance (power) angle is shown in EPRM Eq. 17.59 and Eq. 17.60.

$$P = S \cos \phi \qquad \textit{17.59}$$

$$Q = S \sin \phi \qquad \textit{17.60}$$

The apparent power per phase is shown in EPRM Eq. 33.37.

$$S = VI = \frac{P_p}{\cos \theta} \qquad \textit{33.37}$$

Apparent Power: Phase and Line Version

Handbook: **3-Phase Circuits**

$$|S| = 3V_p I_p = \sqrt{3}\, V_L I_L$$

This equation for apparent power ignores the angular relationship, as is often the case, and incorporates the phase and line values, which are valid regardless of whether a delta or wye connection is being utilized. See EPRM Sec. 24.8, Sec. 24.9, and Sec. 29.3 for an in-depth explanation.

Apparent Power: Combination Version

Handbook: **3-Phase Circuits**

$$S = 3V_P I_P^* = \sqrt{3}\, V_L I_L (\cos \theta_P + j \sin \theta_P)$$

EPRM: Sec. 29.2

$$\begin{aligned} \mathbf{S} &= \mathbf{VI}^* \\ &= P + jQ \qquad\qquad \textit{29.1} \\ &= VI \cos \theta + jVI \sin \theta \end{aligned}$$

Given that the power angle is equal to the power factor angle and that $\pm \cos \phi = |\pm \cos \phi|$, a positive or negative sign cannot be used to indicate leading or lagging. This property must be indicated using words.

EPRM Eq. 29.1 is the single-phase version of the three-phase equation shown in the *Handbook*.

The relationship in EPRM Eq. 24.60 comes from the power triangle, which always shows real power as the reference, reactive power displaced 90° from the real power, and the apparent power connecting the two.

$$S^2 = P^2 + Q^2 \qquad \textit{24.60}$$

EPRM Fig. 17.11 shows the power triangle.

Figure 17.11 *Power Triangle*

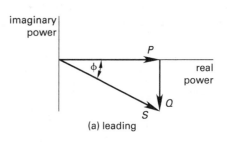

(a) leading

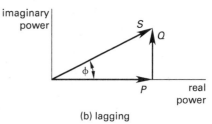

(b) lagging

A useful memory aid is "ELI the ICE man". The voltage, E, leads current, I, in an inductive, L, circuit. Current, I, leads voltage, E, in a capacitive, C, circuit.

3-Phase, Wye-Connected Source or Load with Line-to-Neutral Voltages and a Positive (ABC) Phase Sequence

Handbook: **3-Phase Circuits**

EPRM: Sec. 24.5

$$V_{\text{an}} = V_P \angle 0°$$
$$V_{\text{bn}} = V_P \angle -120°$$
$$V_{\text{cn}} = V_P \angle 120°$$

Though differing in subscripts, the *Handbook* equations and EPRM Eq. 24.16, Eq. 24.17, and Eq. 24.18 are equivalent. Note that $+120°$ and $-240°$ are equivalent angles.

Line-to-Line Voltages for a 3-Phase, Wye-Connected Source or Load with Line-to-Neutral Voltages and a Positive (ABC) Phase Sequence

Handbook: 3-Phase Circuits

EPRM: Sec. 24.5

$$V_{ab} = \sqrt{3}\ V_P \angle 30°$$
$$V_{bc} = \sqrt{3}\ V_P \angle -90°$$
$$V_{ca} = \sqrt{3}\ V_P \angle 150°$$

Though differing in subscripts, the *Handbook* equations and EPRM Eq. 24.22, Eq. 24.23, and Eq. 24.24 are equivalent. Note that $+150°$ and $-210°$ are equivalent angles.

Relationships for Converting Between Equivalent Delta and Wye Connections

Handbook: 3-Phase Circuits

EPRM: Sec. 29.3

$$Z_A = \frac{Z_1 Z_2 + Z_1 Z_3 + Z_2 Z_3}{Z_3}$$
$$Z_B = \frac{Z_1 Z_2 + Z_1 Z_3 + Z_2 Z_3}{Z_1} \quad [Y \rightarrow \Delta]$$
$$Z_C = \frac{Z_1 Z_2 + Z_1 Z_3 + Z_2 Z_3}{Z_2}$$

$$Z_1 = \frac{Z_A Z_C}{Z_A + Z_B + Z_C}$$
$$Z_2 = \frac{Z_A Z_B}{Z_A + Z_B + Z_C} \quad [\Delta \rightarrow Y]$$
$$Z_3 = \frac{Z_B Z_C}{Z_A + Z_B + Z_C}$$

Loads may be converted between equivalent delta and wye connections. The use of subscripts for delta and wye loads differ in the *Handbook* and EPRM. See EPRM Fig. 29.4 for additional clarity.

2. SYMMETRICAL COMPONENTS

Symmetrical components, which represent balanced phasors, greatly simplify calculations. Any set of unbalanced phasors can be expressed as a set of symmetrically balanced phasors, thus making all the equations commonly used in electrical engineering applicable. Such phasors are known as zero-, positive-, and negative-sequence phasors.

Knowledge Area Overview

Key concepts: These key concepts are important for answering exam questions in knowledge area 2.A.2, Symmetrical Components.

- how symmetrical components are used in unsymmetrical fault analysis
- calculations of symmetrical components from phase quantities
- calculations of phase quantities from symmetrical components

PE Power Reference Manual **(EPRM):** Study these sections in EPRM that either relate directly to this knowledge area or provide background information.

- Section 26.7: Fault Analysis: Symmetrical
- Section 26.8: Fault Analysis: Unsymmetrical
- Section 29.7: Sequence Components
- Section 29.8: Symmetrical Components

NCEES Handbook: To prepare for this knowledge area, familiarize yourself with this section in the *Handbook*.

- Symmetrical Components

The following equations are relevant for knowledge area 2.A.2, Symmetrical Components.

Resolve a Set of Unsymmetrical Phasors into a Set of 3-Phase Symmetrical Phasors

Handbook: Symmetrical Components

$$\begin{bmatrix} \mathbf{V}_{a0} \\ \mathbf{V}_{a1} \\ \mathbf{V}_{a2} \end{bmatrix} = \frac{1}{3} \begin{bmatrix} 1 & 1 & 1 \\ 1 & \mathbf{a} & \mathbf{a}^2 \\ 1 & \mathbf{a}^2 & \mathbf{a} \end{bmatrix} \begin{bmatrix} \mathbf{V}_a \\ \mathbf{V}_b \\ \mathbf{V}_c \end{bmatrix}$$

EPRM: Sec. 26.8

$$V_{a0} = \tfrac{1}{3}(V_A + V_B + V_C) \hspace{2cm} 26.27$$

$$V_{a1} = \tfrac{1}{3}(V_A + a V_B + a^2 V_C) \hspace{1.5cm} 26.28$$

$$V_{a2} = \tfrac{1}{3}(V_A + a^2 V_B + a V_C) \hspace{1.5cm} 26.29$$

The *Handbook* and EPRM equations are equivalent. The *Handbook* equations are shown in matrix form, while EPRM shows the resulting individual equations.

EPRM Fig. 26.11 and Fig. 29.8 show the relationships between the sequences and how any unsymmetrical phasor can be constructed.

Construct a Set of Unsymmetrical Phasors from a Set of 3-Phase Symmetrical Phasors

Handbook: Symmetrical Components

$$\begin{bmatrix} \mathbf{V}_a \\ \mathbf{V}_b \\ \mathbf{V}_c \end{bmatrix} = \begin{bmatrix} 1 & 1 & 1 \\ 1 & \mathbf{a}^2 & \mathbf{a} \\ 1 & \mathbf{a} & \mathbf{a}^2 \end{bmatrix} \begin{bmatrix} \mathbf{V}_{a0} \\ \mathbf{V}_{a1} \\ \mathbf{V}_{a2} \end{bmatrix}$$

EPRM: Sec. 26.8

$$V_A = V_{a0} + V_{a1} + V_{a2} \qquad 26.24$$

$$V_B = V_{a0} + a^2 V_{a1} + a V_{a2} \qquad 26.25$$

$$V_C = V_{a0} + a V_{a1} + a^2 V_{a2} \qquad 26.26$$

The *Handbook* and EPRM equations are equivalent. The *Handbook* equations are shown in matrix form. EPRM shows the resulting individual equations.

3. PER UNIT SYSTEM

The per-unit system is meant to simplify calculations. All calculations are related to a chosen "base". In three-phase systems the bases chosen are usually voltage and total apparent power. From these two, the base current and impedance can be calculated.

Knowledge Area Overview

Key concepts: These key concepts are important for answering exam questions in knowledge area 2.A.3, Per Unit System.

- the per-unit system and its advantages

- three-phase, single-phase, and per-phase calculations

- base values

- single-phase and three-phase power as a per-unit value

PE Power Reference Manual **(EPRM):** Study these sections in EPRM that either relate directly to this knowledge area or provide background information.

- Section 24.11: Per-Unit Calculations

- Section 26.9: Fault Analysis: The MVA Method

- Section 29.6: Per-Unit System

NCEES Handbook: To prepare for this knowledge area, familiarize yourself with this section and these tables in the *Handbook*.

- Per Unit System

The following equations and tables are relevant for knowledge area 2.A.3, Per Unit System.

Per-Unit Equations: Summary

Handbook: Per Unit System

EPRM: Sec. 29.6

See EPRM Table 29.1 for a summary of the single-phase and three-phase equations that are provided in the *Handbook*.

Base Current and Impedance for Single-Phase Systems

Handbook: Per Unit System

$$\text{base current, A} = \frac{\text{base kVA}_{1\phi}}{\text{base voltage, kV}_{\text{LN}}}$$

$$\text{base impedance, } \Omega = \frac{\text{base voltage, kV}_{\text{LN}}}{\text{base current, A}}$$

$$= \frac{(\text{base voltage, kV}_{\text{LN}})^2}{\text{base MVA}_{1\phi}}$$

EPRM: Sec. 24.11

$$I_{\text{base}} = \frac{S_{\text{base}}}{V_{\text{base}}} = \frac{S_p}{V_p} = \left(\frac{S}{\sqrt{3}\, V} \right)_{3\phi} \qquad 24.43$$

$$Z_{\text{base}} = \frac{V_{\text{base}}}{I_{\text{base}}} = \frac{V_p^2}{S_p}$$

$$= \left(\frac{V^2}{S} \right)_{3\phi} \qquad 24.44$$

$$= \left(\frac{V}{\sqrt{3}\, I} \right)_{3\phi}$$

The *Handbook* and EPRM equations shown are equivalent. However, the equations in EPRM do not indicate the specific units used in the *Handbook* equations.

Base Current and Impedance for 3-Phase Systems

Handbook: Per Unit System

EPRM: Sec. 24.11

$$\text{base current, A} = \frac{\text{base kVA}_{3\phi}}{\sqrt{3}\,(\text{base voltage, kV}_{\text{LL}})}$$

$$\text{base impedance, } \Omega = \frac{\text{base voltage, kV}_{\text{LN}}}{\text{base current, A}}$$

$$= \frac{(\text{base voltage, kV}_{\text{LL}})^2}{\text{base MVA}_{3\phi}}$$

The *Handbook* and EPRM equations shown are equivalent. However, the equations in EPRM do not indicate the specific units used in the *Handbook* equations. From EPRM Eq. 24.43 and Eq. 24.44,

$$I_{\text{base}} = \frac{S_{\text{base}}}{V_{\text{base}}} = \left(\frac{S}{\sqrt{3}\,V}\right)_{3\phi}$$

$$Z_{\text{base}} = \frac{V_{\text{base}}}{I_{\text{base}}} = \left(\frac{V}{\sqrt{3}\,I}\right)_{3\phi}$$

Change of Base for Per-Unit Impedance

Handbook: Per Unit System

$$Z_{\text{new}} = Z_{\text{old}}\left(\frac{\text{base kV}_{\text{old}}}{\text{base kV}_{\text{new}}}\right)^2\left(\frac{\text{base kVA}_{\text{new}}}{\text{base kVA}_{\text{old}}}\right)$$

EPRM: Sec. 24.11

$$X_{\text{pu,new}} = X_{\text{pu,old}}\left(\frac{X_{\text{base,old}}}{X_{\text{base,new}}}\right) \qquad 24.53$$

The equation shown in the *Handbook* can only be used on impedance values. EPRM Eq. 24.53 applies to the general case.

The per-unit impedance values on either side of a transformer are identical. An example is shown in EPRM *Table for Solution 24.4*.

Because the per-unit impedance is the same on either side of a transformer, a single-line diagram can be drawn and calculations made based on such a diagram, greatly simplifying three-phase analysis.

4. PHASOR DIAGRAMS

The topic of phasors overlaps several other topics. A phasor is a line used to represent a complex electrical quantity as a vector. The phase of an alternating quantity at any instant in time can be represented by a phasor diagram. An understanding of complex numbers is the intended focus of this review.

Knowledge Area Overview

Key concepts: These key concepts are important for answering exam questions in knowledge area 2.A.4, Phasor Diagrams.

- leading/lagging circuits
- calculation of voltages and currents for different impedances

PE Power Reference Manual (**EPRM**): Study these sections in EPRM that either relate directly to this knowledge area or provide background information.

- Section 2.18: Complex Numbers
- Section 2.19: Operations on Complex Numbers
- Chapter 17: AC Circuit Fundamentals

NCEES Handbook: To prepare for this knowledge area, familiarize yourself with this section in the *Handbook*.

- Phasor Transforms of Sinusoids

The following equations and tables are relevant for knowledge area 2.A.4, Phasor Diagrams.

Phasors

Handbook: Phasor Transforms of Sinusoids

$$P\big(V_{\max}\cos(\omega t + \phi)\big) = V_{\text{rms}}\angle\phi = V$$

$$P\big(I_{\max}\cos(\omega t + \theta)\big) = I_{\text{rms}}\angle\theta = I$$

Phasor form is a complex quantity with a magnitude and an angle. The frequency is implied in phasor form.

EPRM Table 17.1 shows the relationship between different circuit elements.

Note that vectors generally do not rotate; however, phasors do rotate. The rotation is often constant and so can be represented in any of the steady-state forms shown. In fact, the rotating portion is normally assumed and not shown at all. For a full explanation, see EPRM Chap. 17.

Table 17.1 *Characteristics of Resistors, Capacitors, and Inductors*

	resistor	capacitor	inductor
value	$R\ (\Omega)$	$C\ (\text{F})$	$L\ (\text{H})$
reactance, X	0	$\dfrac{-1}{\omega C}$	ωL
rectangular impedance, $\mathbf{Z}$	$R + j0$	$0 - \dfrac{j}{\omega C}$	$0 + j\omega L$
phasor impedance, $\mathbf{Z}$	$R\angle 0°$	$\dfrac{1}{\omega C}\angle{-90°}$	$\omega L\angle 90°$
phase	in-phase	leading	lagging
rectangular admittance, $\mathbf{Y}$	$\dfrac{1}{R} + j0$	$0 + j\omega C$	$0 - \dfrac{j}{\omega L}$
phasor admittance, $\mathbf{Y}$	$\dfrac{1}{R}\angle 0°$	$\omega C\angle 90°$	$\dfrac{1}{\omega L}\angle{-90°}$

Characteristics of Circuit Elements

Handbook: Phasor Transforms of Sinusoids

EPRM: Sec. 17.13–17.15

$$Z = \frac{V}{I}$$

$$Z_R = R$$

$$Z_C = \frac{1}{j\omega C} = jX_C$$

$$Z_L = j\omega L = jX_L$$

The impedance is the ratio of the phasor voltage to the phasor current. The *Handbook* equations shown are equivalent to EPRM Eq. 19.5, Eq. 17.34, Eq. 17.35, and Eq. 17.37, respectively.

Complex Number Representation

Handbook: Phasor Transforms of Sinusoids

EPRM: Sec. 17.12

$$P\big(V_{\max}\cos(\omega t + \phi)\big) = V_{\text{rms}}\angle\phi = V$$
$$P\big(I_{\max}\cos(\omega t + \theta)\big) = I_{\text{rms}}\angle\theta = I$$

Phasors are plotted in the complex plane. See EPRM Table 17.3 for a summary of the properties of complex numbers.

5. SINGLE-PHASE CIRCUITS

A single-phase AC circuit uses a power wire and a neutral wire to provide power to a load. Single-phase AC circuits have a wide variety of applications in household usage, are less complex than three-phase designs, and are less expensive.

Knowledge Area Overview

Key concepts: These key concepts are important for answering exam questions in knowledge area 2.A.5, Single-Phase Circuits.

- function of capacitors and the factors affecting capacitance

- equivalent capacitance in parallel and in series

- energy storage in a capacitor

- function of inductors and the factors affecting inductance

- inductances in parallel and in series

- conditions under which maximum power transfer can occur

PE Power Reference Manual (**EPRM**): Study these sections in EPRM that either relate directly to this knowledge area or provide background information.

- Section 4.9: Functions of Related Angles

- Section 7.17: Current

- Section 17.2: Voltage

- Section 17.3: Current

- Section 17.4: Impedance

- Section 17.7: Average Value

- Section 17.8: Root-Mean-Square Value

- Section 17.13: Resistors

- Section 17.14: Capacitors

- Section 17.15: Inductors

- Section 17.18: Power

- Section 17.19: Real Power and the Power Factor

- Section 17.20: Reactive Power

- Section 23.11: Harmonics

- Chapter 24: Three-Phase Electricity and Power

- Section 26.8: Fault Analysis: Unsymmetrical

- Section 27.3: Transformer Capacity

- Section 27.5: Three-Phase Transformer Configurations

- Section 27.11: Transformer Parallel Operation

- Section 27.15: Open-Delta Transformers

- Section 28.5: Internal Inductance

- Section 28.7: Single-Phase Inductance

- Section 28.8: Single-Phase Capacitance

- Section 29.2: Power Flow

NCEES Handbook: To prepare for this knowledge area, familiarize yourself with these sections in the *Handbook.*

- Single-Phase Circuits: AC Circuits
- Average Value
- Effective or RMS Values
- Sine-Cosine Relations and Trigonometric Identities
- Phasor Transforms of Sinusoids
- Complex Power

The following equations figure, and tables are relevant for knowledge area 2.A.5, Single-Phase Circuits.

Sinusoidal Voltage

Handbook: Single-Phase Circuits: AC Circuits

EPRM: Sec. 17.2

$$f = \frac{1}{T} = \frac{\omega}{2\pi}$$

Frequency, *f*, is in cycles per second. Recall that a cycle is not a unit. The period, *T*, is in seconds. The angular frequency units are in radians per second. One radian is approximately 57.3°.

The natural or fundamental frequency of a waveform is determined using EPRM Eq. 6.48. The frequency depends upon the period.

$$\omega = \frac{2\pi}{T} \qquad 6.48$$

Periodic Waveform (Either Voltage or Current)

Handbook: Average Value

$$X_{\text{ave}} = \frac{1}{T} \int_0^T x(t)\, dt$$

EPRM: Sec. 17.7

$$V_{\text{ave}} = \frac{1}{2\pi} \int_0^{2\pi} v(\theta)\, d\theta \qquad 17.16$$
$$= \frac{1}{T} \int_0^T v(t)\, dt$$

The *Handbook* and EPRM equations are equivalent. The *X* used in the *Handbook* equation is a generic variable that can represent current or voltage.

The period for a cycle is 2π radians.

Average Value of a Full-Wave Rectified Sinusoid

Handbook: Average Value

$$X_{\text{ave}} = \frac{2X_{\text{max}}}{\pi}$$

EPRM: Sec. 17.7

$$V_{\text{ave}} = \frac{1}{\pi} \int_0^{\pi} v(\theta)\, d\theta \qquad 17.17$$
$$= \frac{2V_m}{\pi} \quad \text{[rectified sinusoid]}$$

The *Handbook* and EPRM equations are equivalent. The *Handbook* equation uses a generic variable, *X*, instead of voltage, *V*.

RMS or Effective Value for a Periodic Waveform

Handbook: Effective or RMS Values

$$X_{\text{eff}} = X_{\text{rms}} = \left(\frac{1}{T} \int_0^T x^2(t)\, dt \right)^{1/2}$$

EPRM: Sec. 17.8

$$f_{\text{rms}}^2 = \frac{1}{T} \int_{t_1}^{t_1 + T} f^2(t)\, dt \qquad 17.18$$

The *Handbook* and EPRM equations are equivalent. The *Handbook* uses *X* for a generic variable and takes the square root of the right side of the equation.

EPRM Eq. 17.19 uses voltage and takes the square root of the right side of the equation, which more closely matches the *Handbook* equation.

$$V = V_{\text{eff}}$$
$$= V_{\text{rms}}$$
$$= \sqrt{\frac{1}{T} \int_0^T v^2(t)\, dt} \qquad 17.19$$
$$= \sqrt{\frac{1}{2\pi} \int_0^{2\pi} v^2(\theta)\, d\theta}$$

RMS or Effective Value for a Sinusoidal Waveform and Full-Wave Rectified Sine Wave

Handbook: Effective or RMS Values

$$X_{\text{eff}} = X_{\text{rms}} = \frac{X_{\text{max}}}{\sqrt{2}}$$

EPRM Table 17.2 indicates the full-wave rectified sinusoid relationship in terms of a ratio. It is equivalent to the *Handbook* equation shown, with voltage, V, substituting for X.

RMS or Effective Value for a Half-Wave Rectified Sine Wave

Handbook: Effective or RMS Values

$$X_{\text{eff}} = X_{\text{rms}} = \frac{X_{\text{max}}}{2}$$

EPRM Table 17.2 shows the half-wave rectified sinusoid relationship in terms of a ratio. It is equivalent to the *Handbook* equation, with voltage, V, substituting for X.

RMS or Effective Value for a Periodic Signal

Handbook: Effective or RMS Values

EPRM: Sec 23.11

$$X_{\text{rms}} = \sqrt{X_{\text{DC}}^2 + \sum_{n=1}^{\infty} X_n^2}$$

The *Handbook* equation represents the impact of harmonics. X_{DC} represents the DC component of the signal, if any. The summation term represents the rms value of the nth harmonic. The first harmonic is called the fundamental harmonic. In the United States, this is the 60 Hz signal.

The harmonic name, frequency, and sequence are shown in EPRM Table 23.1. The positive sequence tends to result in overheating. The negative sequence weakens motor magnetic fields, especially in induction motors. The zero sequence can result in increased currents in the neutral of a four-wire system.

Sine-Cosine Relations and Trigonometric Identities

Handbook: Sine-Cosine Relations and Trigonometric Identities

$$\cos(\omega t) = \sin\left(\omega t + \frac{\pi}{2}\right) = -\sin\left(\omega t - \frac{\pi}{2}\right)$$

$$\sin(\omega t) = \cos\left(\frac{\omega t - \pi}{2}\right) = -\cos\left(\frac{\omega t + \pi}{2}\right)$$

A summary of sine-cosine relations and related angles is shown in EPRM Table 4.2.

Table 4.2 *Functions of Related Angles*

$f(\theta)$	$-\theta$	$90° - \theta$	$90° + \theta$	$180° - \theta$	$180° + \theta$
sin	$-\sin\theta$	$\cos\theta$	$\cos\theta$	$\sin\theta$	$-\sin\theta$
cos	$\cos\theta$	$\sin\theta$	$-\sin\theta$	$-\cos\theta$	$-\cos\theta$
tan	$-\tan\theta$	$\cot\theta$	$-\cot\theta$	$-\tan\theta$	$\tan\theta$

Phasor Transforms of Sinusoids

Handbook: Phasor Transforms of Sinusoids

$$P\left(V_{\text{max}}\cos(\omega t + \phi)\right) = V_{\text{rms}}\angle\phi = \mathbf{V}$$

$$P\left(I_{\text{max}}\cos(\omega t + \theta)\right) = I_{\text{rms}}\angle\theta = \mathbf{I}$$

EPRM: Sec. 17.2

trigonometric form:

$$V_m \sin(\omega t + \theta)$$

exponential form:

$$V_m e^{j\theta}$$

polar or phasor form:

$$V_m \angle\theta$$

rectangular form:

$$V_r + jV_i$$

Electrical quantities can be represented in many forms. The trigonometric form arguably displays the most information. The exponential information form exhibits the rotational information. The polar or phasor form is the easiest to use for performing calculations. Refer to the Steinmetz algorithm from EPRM Fig. 17.8.

Figure 17.8 *Steinmetz Algorithm Steps*

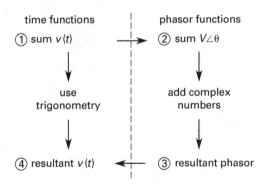

The rectangular form separates the real and reactive components.

Impedance of a Circuit Element

Handbook: Phasor Transforms of Sinusoids

$$Z = \frac{V}{I}$$

The impedance shown here is in bold and indicates phasor quantities. Angles are associated with each quantity.

EPRM: Sec. 7.17

$$V = IR \qquad\qquad \textit{7.32}$$

EPRM Eq. 7.32 is the most basic form of Ohm's law and is equivalent to the *Handbook* equation. The angular relationship is not shown in the EPRM equation, indicating that the magnitude is the desired quantity, or that the load is purely resistive and the angle between the voltage and current is zero degrees.

Resistive Impedance

Handbook: Phasor Transforms of Sinusoids

$$Z_R = R$$

EPRM: Sec. 17.13

$$\mathbf{Z}_R = R\angle 0 = R + j0 \qquad\qquad \textit{17.34}$$

An ideal resistor has no reactive component, hence the zero-degree angle. Therefore, the angle is often not shown.

A summary of the characteristics of resistors, capacitors, and inductors is shown in EPRM Table 17.1.

Table 17.1 *Characteristics of Resistors, Capacitors, and Inductors*

	resistor	capacitor	inductor
value	$R\,(\Omega)$	$C\,(\text{F})$	$L\,(\text{H})$
reactance, X	0	$\dfrac{-1}{\omega C}$	ωL
rectangular impedance, $\mathbf{Z}$	$R + j0$	$0 - \dfrac{j}{\omega C}$	$0 + j\omega L$
phasor impedance, $\mathbf{Z}$	$R\angle 0°$	$\dfrac{1}{\omega C}\angle -90°$	$\omega L\angle 90°$
phase	in-phase	leading	lagging
rectangular admittance, $\mathbf{Y}$	$\dfrac{1}{R} + j0$	$0 + j\omega C$	$0 - \dfrac{j}{\omega L}$
phasor admittance, $\mathbf{Y}$	$\dfrac{1}{R}\angle 0°$	$\omega C\angle 90°$	$\dfrac{1}{\omega L}\angle -90°$

Capacitive Impedance

Handbook: Phasor Transforms of Sinusoids

$$Z_C = \frac{1}{j\omega C} = jX_C$$

EPRM: Sec. 17.14

$$\mathbf{Z}_C = X_C\angle -90° = 0 + jX_C \qquad\qquad \textit{17.35}$$

The capacitive reactance has a negative impedance, hence the $-90°$ angle. The impedance comes late in a given cycle, meaning the current comes first, followed by the resistance to the current (due to the buildup of electrons on the plates of the capacitor).

Inductive Impedance

Handbook: Phasor Transforms of Sinusoids

$$Z_L = j\omega L = jX_L$$

EPRM: Sec. 17.15

$$\mathbf{Z}_L = X_L\angle 90° = 0 + jX_L \qquad\qquad \textit{17.37}$$

The inductive reactance has a positive impedance, hence the $+90°$ angle. The impedance comes early in a given cycle, meaning the impedance comes first and the current comes after the voltage builds up the magnetic field. Once the magnetic field builds up, the current is allowed to pass.

Capacitive Reactance, X_C

Handbook: Phasor Transforms of Sinusoids

EPRM: Sec. 17.14

$$X_C = -\frac{1}{\omega C}$$

The capacitive reactance is negative. The impact of capacitive reactance comes later in a given cycle, meaning the current occurs first, followed by resistance to the current (due to the buildup of electrons on the plates of the capacitor). It is standard practice to not show the negative sign in the capacitive reactance term. Do not expect to see it explicitly.

In the United States, the fundamental frequency is 60 Hz. The equation in EPRM Table 33.1 shows the angular velocity for the 60 Hz potential.

$$\omega = 2\pi f$$

Inductive Reactance, X_L

Handbook: **Phasor Transforms of Sinusoids**

EPRM: Sec. 17.15

$$X_L = \omega L$$

The inductive reactance is positive. The impact of inductive reactance comes earlier in a given cycle. Impedance comes first, and current comes last after the voltage builds up the magnetic field. Once the magnetic field builds up, the current is allowed to pass.

In the United States, the fundamental frequency is 60 Hz. The equation in EPRM Table 33.1 shows the angular velocity for the 60 Hz potential.

$$\omega = 2\pi f$$

Real Power

Handbook: **Complex Power**

$$P = \tfrac{1}{2} V_{\max} I_{\max} \cos\theta$$
$$= V_{\mathrm{rms}} I_{\mathrm{rms}} \cos\theta$$

EPRM: Sec. 17.19

$$P_{\mathrm{ave}} = \left(\frac{I_m V_m}{2} \right) \cos\phi_{\mathrm{pf}} \qquad 17.52$$
$$= I_{\mathrm{rms}} V_{\mathrm{rms}} \cos\phi_{\mathrm{pf}}$$

The *Handbook* and EPRM equations are equivalent. The following terms are also equivalent.

$$\cos\theta = \cos\phi_{\mathrm{pf}} = \mathrm{pf}$$
$$P_{\mathrm{ave}} = P$$

Power Factor

Handbook: **Complex Power**

$$\mathrm{pf} = \cos\theta$$

EPRM: Sec. 17.19

$$P = I V \cos\phi_{\mathrm{pf}} \qquad 17.53$$
$$= I V \,\mathrm{pf}$$

EPRM Eq. 17.53 shows the simplification from the cosine to the power factor.

Often, the real and apparent power are known values. From these known values, the power factor may be calculated.

The reactance and resistance of a given circuit are related via the power factor. This is shown in EPRM Eq. 17.61 and Eq. 17.62.

$$\mathrm{pf} = \frac{P}{S} \qquad 17.61$$

$$\frac{X}{R} = \tan(\arccos\mathrm{pf}) \qquad 17.62$$

Reactive Power

Handbook: **Complex Power**

$$Q = \tfrac{1}{2} V_{\max} I_{\max} \sin\theta$$
$$= V_{\mathrm{rms}} I_{\mathrm{rms}} \sin\theta$$

EPRM: Sec. 17.20

$$Q = I V \sin\phi_{\mathrm{pf}} \qquad 17.56$$

The *Handbook* and EPRM equations are equivalent. The subscript "rms" is usually assumed and is not shown in the EPRM equation.

Complex Power

Handbook: **Complex Power, Complex Power Triangle (Inductive Load)**

EPRM: Sec. 29.2

$$S = VI^* = P + jQ$$

$$S = VI$$

$$Q = VI \sin\theta$$

$$P = VI \cos\theta$$

Complex power, **S** or S, is in units of volt-amperes. **I*** is the complex conjugate of the phasor current.

6. DC CIRCUITS

A DC circuit is one in which current flows unidirectionally only. While current may flow in one direction, it may be variable, or it might be described as pulsing, but it will not change direction. Many of the concepts in this knowledge area are not covered in the *NCEES Handbook* but are nevertheless important to know for the PE exam. The relevant *PE Power Reference Manual* sections are provided and are recommended for study.

Knowledge Area Overview

Key concepts: These key concepts are important for answering exam questions in knowledge area 2.A.6, DC Circuits.

- basic DC circuit element symbols
- current, voltage, and resistance
- Ohm's law and Kirchhoff's law
- Thevenin equivalent analysis
- Norton equivalent analysis
- equivalent resistance for resistors in series and in parallel
- node circuit analysis
- loop circuit analysis

PE Power Reference Manual **(EPRM):** Study these sections in EPRM that either relate directly to this knowledge area or provide background information.

- Section 7.17: Current
- Chapter 16: DC Circuit Fundamentals
- Section 19.9: Resistance
- Section 19.10: Capacitance
- Section 19.11: Inductance
- Section 19.18: Thevenin's Theorem
- Section 19.19: Norton's Theorem
- Section 19.27: Node Analysis
- Section 19.28: Determination of Method
- Section 19.30: Steady-State and Transient Impedance Analysis
- Section 20.1: Fundamentals
- Section 28.2: DC Resistance

NCEES Handbook: To prepare for this knowledge area, familiarize yourself with these sections in the *Handbook.*

- DC Circuits
- Transmission Line Parameters

The following equations and table are relevant for knowledge area 2.A.6, DC Circuits.

RC Circuit Transient: Capacitor Voltage

Handbook: DC Circuits

$$v_C(t) = v_C(0)e^{-t/RC} + V\left(1 - e^{-t/RC}\right)$$

EPRM: Table 20.1

$$v_c(t) = V_0 + (V_{\text{bat}} - V_0)$$
$$\times\left(1 - e^{-N}\right)$$

The *Handbook* equation represents an RC transient on a DC circuit with the capacitor at an initial charge of $v_C(0)$. This is the equivalent of the equation for capacitor voltage from EPRM Table 20.1; only the symbology differs.

RC Circuit Transient: Current

Handbook: DC Circuits

$$i(t) = \left(\frac{V - v_C(0)}{R}\right)e^{-t/RC}$$

EPRM: Table 20.1

$$i(t) = \left(\frac{V_{\text{bat}} - V_0}{R}\right)e^{-N}$$

The *Handbook* equation represents an RC transient on a DC circuit with the capacitor at an initial charge of $v_C(0)$. This is the equivalent of the equation shown for current from EPRM Table 20.1; only the symbology differs.

RC Circuit Transient: Resistor Voltage

Handbook: DC Circuits

$$v_R(t) = i(t)R = (V - v_C(0))e^{-t/RC}$$

EPRM: Table 20.1

$$v_R(t) = i(t)R$$
$$= (V_{\text{bat}} - V_0)e^{-N}$$

The *Handbook* displays an equation that represents an RC transient on a DC circuit with the capacitor at an initial charge of $v_C(0)$. This is the equivalent of the equation shown for resistor voltage from EPRM Table 20.1; only the symbology differs.

RL Circuit Transient: Current

Handbook: DC Circuits

$$i(t) = i(0)e^{-Rt/L} + \left(\frac{V}{R}\right)\left(1 - e^{-Rt/L}\right)$$

EPRM: Table 20.1

$$i(t) = I_0 e^{-N}$$
$$+ \left(\frac{V_{\text{bat}}}{R} \right) \left(1 - e^{-N} \right)$$

The *Handbook* equation represents an RL transient on a DC circuit with the inductor initial current of $i(0)$. This is the equivalent of the equation shown for current from EPRM Table 20.1; only the symbology differs.

RL Transient: Resistor Voltage

Handbook: DC Circuits

$$v_R(t) = i(t)R = \left(V - v_C(0) \right) e^{-t/RC}$$

EPRM: Table 20.1

$$v_R(t) = i(t)R$$
$$= I_0 R e^{-N} + V_{\text{bat}} (1 - e^{-N})$$

The *Handbook* equation represents a RL transient on a DC circuit with the inductor initial current of $i(t)$. This is the equivalent of the equation showing resistor voltage from EPRM Table 20.1; only the symbology differs.

RL Transient: Inductor Voltage

Handbook: DC Circuits

$$v_L(t) = L \frac{di}{dt} = -i(0) R e^{-Rt/L} + V e^{-Rt/L}$$

EPRM: Table 20.1

$$v_L(t) = -I_0 R e^{-N}$$

The *Handbook* equation represents an RL transient on a DC circuit with the inductor initial current of $i(0)$. This is the equivalent of the equation shown for inductor voltage from EPRM Table 20.1; only the symbology differs.

Resistance

Handbook: Transmission Line Parameters

EPRM: Sec. 16.3

$$R_{\text{DC}} = \frac{\rho l}{A}$$

A is area, l is length, and ρ is resistivity. The *Handbook* equation and EPRM Eq. 16.1 are equivalent.

Equivalent Resistance for Resistors in Series

EPRM: Sec. 16.17

$$R_e = R_1 + R_2 + R_3 + \cdots + R_n \qquad \textit{16.27}$$

For resistors in series, the equivalent resistance is the sum of the individual resistances.

Equivalent Resistance for Resistors in Parallel

EPRM: Sec. 16.18

$$\frac{1}{R_e} = \frac{1}{R_1} + \frac{1}{R_2} + \frac{1}{R_3} + \cdots + \frac{1}{R_n} \qquad \textit{16.31(a)}$$

For resistors in parallel, the reciprocal equivalent resistance is the sum of the reciprocals of the individual resistances.

For two resistors in parallel,

$$R_e = \frac{R_1 R_2}{R_1 + R_2}$$

The equation for two resistors derives from general case.

Ohm's Law

EPRM: Sec. 16.5

$$V = IR \qquad \textit{16.11}$$

Ohm's law is the most fundamental and important electrical law.

Kirchhoff's Laws

EPRM: Sec. 16.15–Sec. 16.16

$$\sum_{\text{loop}} \text{voltage rises} = \sum_{\text{loop}} \text{voltage drops} \qquad \textit{16.24}$$

$$\sum_{\text{node}} \text{currents in} = \sum_{\text{node}} \text{currents out} \qquad \textit{16.25}$$

Kirchhoff's laws are really conservation of energy equations. It does not matter whether the elements are purely resistive or reactive; Kirchhoff's voltage law (KVL) and Kirchhoff's current law (KCL) still apply.

Thevenin Equivalent Analysis

EPRM: Sec. 16.24

$$V_{\text{Th}} = V_{\text{oc}} \hspace{2cm} 16.40$$

$$R_{\text{Th}} = \frac{V_{\text{oc}}}{I_{\text{sc}}} \hspace{2cm} 16.41$$

The Thevenin equivalent voltage and resistance are calculated using EPRM Eq. 16.40 and Eq. 16.41.

Review EPRM Sec. 16.24 for the steps used to determine the Thevenin equivalent.

Norton Equivalent Analysis

EPRM: Sec. 16.25

$$I_N = I_{\text{sc}} \hspace{2cm} 16.42$$

$$R_N = \frac{V_{\text{oc}}}{I_{\text{sc}}} \hspace{2cm} 16.43$$

The Norton equivalent current and resistance are calculated using EPRM Eq. 16.42 and Eq. 16.43.

Norton and Thevenin resistances are equal.

The Norton equivalent replaces a network with a current source (which produces constant current) in parallel with a resistor. The open-circuit voltage is measured with all voltage sources shorted.

Node-Voltage Method

EPRM: Sec. 16.27

The node-voltage method is a systematic network-analysis procedure that uses voltages as the unknowns. Review EPRM Sec. 16.27 for the process to write node equations.

Superposition

EPRM: Sec. 16.23

The principle of *superposition* is that the response (that is, the voltage across or current through) of a linear circuit element in a network with multiple independent sources is equal to the response obtained if each source is considered individually and the results are summed.

Review EPRM Sec. 16.23 for the steps involved in determining the desired quantity.

Superposition holds true only for linear systems. If the systems are not linear, then they cannot be solved via superposition.

7. SINGLE-LINE DIAGRAMS

Single-line diagrams are used for load studies, fault analysis, coordination studies, general connection of equipment, and sequence of operations. In the *NCEES Handbook*, the topic of single-line diagrams is limited to a listing of IEEE standard protective device relay numbers.

Knowledge Area Overview

Key concepts: These key concepts are important for answering exam questions in knowledge area 2.A.7, Single-Line Diagrams.

- uses of single-line diagrams
- items that should be on a single-line diagram
- analysis of three-phase systems using single-line diagrams

PE Power Reference Manual **(EPRM):** Study these sections in EPRM that either relate directly to this knowledge area or provide background information.

- Section 24.6: Distribution Systems
- Section 24.11: Per-Unit Calculations
- Section 26.1: Fundamentals
- Section 29.6: Per-Unit System
- Chapter 31: Protection and Safety

NCEES Handbook: To prepare for this knowledge area, familiarize yourself with this section and table in the *Handbook.*

- Single-Line Diagrams

The following equations, figures, and tables are relevant for knowledge area 2.A.7, Single-Line Diagrams.

Single-Line Diagrams

Single-line diagrams are widely used to represent distribution systems in a simplified manner and to ease first-order calculations. A three-phase diagram and the corresponding single-line diagram are shown.

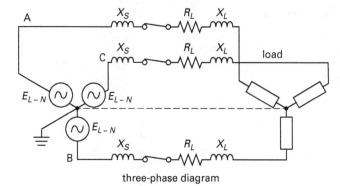

three-phase diagram

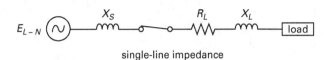

single-line impedance

The following figure is another example of a one-line diagram of a three-phase system. This particular drawing is used to calculate a fault current. Note the clarity and simplification.

The transformer can be represented as a single line because the per-unit impedance is the same on both sides. An example of such a calculation is shown in EPRM *Table for Solution 24.4* (which is unrelated to the one-line diagram shown). Therefore, even though the voltage changes, the per-unit impedance is the same and a three-phase problem becomes a simple Ohm's law calculation.

Device Number Designations

Handbook: Single-Line Diagrams

EPRM: Sec. 31.4

The *Handbook* includes a table of device number designations according to ANSI/IEEE C37.2 for common power system elements. EPRM Table 31.1 also shows an assortment of devices and their functions.

The IEEE Standard contains a great deal of device numbers. The standard would have to be provided for the examinee, or at least the relevant information provided in the problem statement, as this information is not something that requires memorization.

8. DEFINITIONS

apparent power: The combination of real and reactive power, usually without reference to an angle.

complex power: A vector value that is the sum of the reactive power and real power.

delta connection: A three-phase connection in which the three windings are connected end-to-end that are 120° apart from each other electrically.

frequency: The number of cycles per second of current, which depends on the period of the current. The frequency chosen in the United States was 60 Hz, based on minimizing light flicker and the operation of induction motors.

impedance: The complex-valued generalization of resistance; the opposition that a circuit presents to a current when a voltage is applied.

negative-sequence phasors: Three equal-magnitude phasors rotating counterclockwise in the sequence CBA.

per-unit: A per-unit value is any quantity actual value related to a given base.

phase sequence: The sequence in which the phase voltages achieve their maximum positive values. Reference to the zero x-axis. The rotation is counterclockwise. Positive sequence is ABC. Negative sequence is CBA.

phasor: A line used to represent a complex electrical quantity as a vector. The diagrams in which they appear are phasor diagrams. Phasors rotate; the rotation is often constant and so can be represented in a steady-state form.

positive-sequence phasors: Three equal-magnitude phasors rotating counterclockwise in the sequence ABC.

power angle: The phase angle difference between the applied voltage and the generated emf. Also known as torque angle.

power factor: The ratio of real power to apparent power measured in kilovolt-amperes.

reactive power: The product of the rms values of the current and voltage multiplied by the quadrature of the current.

real power: The average power consumed by the resistive elements of a circuit. Also known as true power.

rectifier: An electrical device that converts alternating current (AC), which periodically reverses direction, to direct current (DC).

RC circuit: A circuit with both a resistor, R, and a capacitor, C.

RL circuit: A circuit with both a resistor, R, and an inductor, L.

rms voltage: A synonym for mean power or average power; it is the square root of the time average of the voltage squared.

single-line diagram: Also called a one-line diagram; a simplified notation for representing a three-phase power system.

sinusoid: A sine wave; a continuous wave named after the function of sine.

superposition theorem: For a linear system, the voltage or current in any branch of a bilateral linear circuit having more than one independent source equals the algebraic sum of the voltages or currents caused by each independent source acting alone.

wye connection: A three-phase connection in which all three loads are connected at a single neutral point.

zero-sequence phasors: Three equal-magnitude phasors coincident in phase sequence and rotating counterclockwise.

9. NOMENCLATURE

A	area	m^2
C	capacitance	F
E	generated voltage	V
f	frequency	Hz (cycles/s)
i	instantaneous current	A
I, I	DC or rms current	A
l	length	m
L	inductance	H
P	real power	W
P	phasor	varies
pf	power factor	–
Q	reactive power	VAR
R	resistance	Ω
S	apparent power	VA
t	time	s
T	period	s
v	instantaneous voltage	V
V, V	DC or rms voltage	V
X	reactance	Ω
Z, Z	impedance	Ω

c	phase c
C	capacitive
C	characteristic
C	phase C
cn	voltage c to neutral
e	equivalent
eff	effective
i	imaginary
L	inductive
L	load
l, L	line
LL	line to line
LN	line to neutral
L-N	line to neutral
m	maximum
max	maximum
n	number
N	Norton equivalent
n, N	neutral
oc	open circuit
p, ϕ	phase
pu	per unit
R	resistive
r	real or radial
rms	root mean square
S	secondary
sc	short circuit
Th	Thevenin

Symbols

θ	angle	radians
ρ	resistivity	$\Omega \cdot$m
ϕ	power angle	radians
ϕ_I	current angle	radians
ω	angular frequency	rad/s

Subscripts

0	original
a	phase a
a0	zero sequence
a1	positive sequence
a2	negative sequence
A	phase A
an	voltage a to neutral
ave	average
avg	average
b	phase b
bat	battery
bn	voltage b to neutral
B	phase B

5 Devices and Power Electronic Circuits

Exam specification 2.B, Devices and Power Electronic Circuits, makes up between 8% and 11% of the PE Electrical Power exam (between 6 and 9 questions out of 80).

The organization of this chapter follows the order of knowledge areas given by the NCEES for this exam specification. Each knowledge area is covered in the following numbered sections.

Content in blue refers to the *NCEES Handbook*.

Content in red is additional essential information.

1. BATTERY CHARACTERISTICS AND RATINGS

Battery advancements are developing rapidly with a focus on energy density. The potential difference developed between cells depends upon the chemistry, while the capacity depends upon the amount of material.

Knowledge Area Overview

Key concepts: These key concepts are important for answering exam questions in knowledge area 2.B.1, Battery Characteristics and Ratings.

- electrical properties of batteries

- equivalent circuit model

- battery capacity using Peukert's relation for lead-acid batteries

- discharge power and discharge current of batteries

- different types of batteries

- cell voltages of different types of batteries

- self-discharge rate of batteries

PE Power Reference Manual **(EPRM):** Study these sections in EPRM that either relate directly to this knowledge area or provide background information.

- Section 16.29: Practical Application: Batteries

- Section 19.13: Linear Source Models

- Chapter 25: Batteries, Fuel Cells, and Power Supplies

- Section 31.8: Protection System Elements

NCEES Handbook: To prepare for this knowledge area, familiarize yourself with these sections in the *Handbook*.

- Voltage, Capacity, and Specific Energy of Major Battery Systems—Theoretical and Practical Values

- Standard Potential of a Cell

- Theoretical Capacity of a Cell

- Characteristics of Typical Electrode Materials

- Theoretical Energy

- Discharge Rates

- Relationship of Temperature on Battery Capacity

- Peukert's Relation for Lead-Acid Batteries

The following equations and table are relevant for knowledge area 2.B.1, Battery Characteristics and Ratings.

Standard Potential of a Cell

Handbook: Standard Potential of a Cell

EPRM: Sec. 25.2

$$E^0 = V_{\text{cathode}} - V_{\text{anode}}$$

The standard potential of a cell is represented directly by the materials used as the anode and the cathode. E^0 is the potential difference between the two electrodes (cathode and anode) or the standard cell electromotive force (emf). V_{cathode} and V_{anode} are the cathode and anode potentials, respectively. Potentials are standardized and measured against a hydrogen reaction set as an arbitrary 0 V. Standard reversible potentials are shown in EPRM Table 25.2.

Theoretical Capacity of a Cell

Handbook: Theoretical Capacity of a Cell

EPRM: Sec. 25.2

$$C_{\text{cell}} = \cfrac{1}{\cfrac{1}{\text{EC}_a} + \cfrac{1}{\text{EC}_c}}$$

The *Handbook* equation and EPRM Eq. 25.5 are identical. The electrochemical equivalent, EC, represents the mass of a substance produced or consumed at the electrode by one coulomb of charge (or by one ampere of current passed for one second). See Faraday's law of electrolysis for background information.

The theoretical capacity, C, can be represented in terms of the ECs of the electrodes. The units of C will be consistent with the units used for the anode, EC_a, and the cathode, EC_c.

For example, EPRM Eq. 25.3 and Eq. 25.4 use units of mg/C and g/A·h, respectively.

Theoretical Energy

Handbook: Theoretical Energy

$$\text{energy}_{\text{W·h}} = \text{voltage}_{\text{V}} \times \text{charge}_{\text{A·h}}$$

EPRM: Sec. 25.3

$$E = VQ \qquad\qquad 25.9$$

Energy and battery capacity, which is actually a unit of charge in terms of units (A·h), are related by voltage. The *Handbook* equation is equivalent to EPRM Eq. 25.9, where energy, E, is the product of the voltage, V, and the charge, Q.

Discharge Current

Handbook: Discharge Rates

EPRM: Sec. 25.3

$$I = MC_n$$

The discharge current, I, is related to the multiple or fraction, M, of the battery's C-rate in ampere-hours, C, for the specified capacity time in hours, n.

The C-rate is the rate at which a battery is discharged relative to its maximum capacity. For example, a 1C-rate indicates a battery will discharge its entire capacity in 1 hour. If the rated capacity is 100 A·h, then the discharge rate is 100 A during that hour.

Discharge Power

Handbook: Discharge Rates

EPRM: Sec. 25.3

$$P = ME_n$$

The discharge power in watts, P, is related to the multiple fraction, M, of the rated energy (or E-rate) in watt-hours, E, for a specified capacity time in hours, n. The E-rate is the discharge power, in watts, at which a battery will discharge all its power in one hour. For example, a 1E-rate indicates the discharge power to discharge the entire capacity in 1 hour. The voltage at which a battery is fully discharged is called the cutoff voltage.

Peukert's Equation

Handbook: Peukert's Relation for Lead-Acid Batteries

EPRM: Sec. 25.4

$$C_p = I^k t$$

In general terms, Peukert's equation relates the battery capacity and discharge rate under constant temperature and constant discharge current conditions. C_p is the capacity in units of A·h for a 1.0 A discharge current, I. The superscript k is a dimensionless quantity called the Peukert constant, and t is the discharge time in hours. The value of k varies with the battery, but is within a range of 1.1 to 1.4.

Peukert's Equation: Actual Discharge Time to Rated Discharge Time

Handbook: Peukert's Relation for Lead-Acid Batteries

EPRM: Sec. 25.4

$$t = H\left(\frac{C}{IH}\right)^k$$

Since a 1.0 A discharge rate is uncommon, it is more useful to compute the actual discharge time in hours, t, for the actual discharge current (in amperes) at the manufacturer's specified nominal capacity in ampere-hours, C, and rated discharge period in hours, H.

EPRM Eq. 25.12 is another useful way to write Peukert's law in terms of the effective capacity, It (in A·h), at the discharge rate of I (in A).

$$It = C\left(\frac{C}{IH}\right)^{k-1} \qquad\qquad 25.12$$

Types of Batteries

EPRM: Sec. 25.3

Primary batteries are not rechargeable. Dry cell batteries are primary batteries that contain no free or liquid electrolyte.

Secondary batteries can be recharged. Recharging is done by passing current through the battery in the direction opposite that of the discharge current.

Reserve batteries are those in which an essential component is withheld, making the battery inert and permitting long-term storage.

2. POWER SUPPLIES AND CONVERTERS

A power supply is any source of electrical energy, generally refers to a battery or power line, used to supply an electronic circuit with the proper electric voltages and currents for operation. Generally, converters are electrical devices that convert alternating current to the direct current. Inverters convert direct current to the alternating current. Inverters are used in uninterruptible power supplies (UPS).

Knowledge Area Overview

Key concepts: These key concepts are important for answering exam questions in knowledge area 2.B.2, Power Supplies and Converters.

- full-wave and half-wave rectifiers
- instantaneous voltage as a function of time
- ripple factor of a power supply
- characteristics of alternating waveforms
- operating parameters of silicon-controlled rectifiers
- operating parameters and applications of diodes
- purpose and components of a power supply
- applications of silicon-controlled rectifiers in circuits
- output voltages of DC-DC converters

PE Power Reference Manual **(EPRM):** Study these sections in EPRM that either relate directly to this knowledge area or provide background information.

- Section 25.6: Power Supplies
- Section 25.7: Uninterruptible Power Supplies (UPS)
- Section 27.14: Buck Transformers/Boost Transformers/Autotransformers
- Chapter 33: Rotating AC Machinery

NCEES Handbook: To prepare for this knowledge area, familiarize yourself with these sections and this figure in the *Handbook.*

- Uncontrolled Single-Phase Half-Wave Rectifier
- Controlled Single-Phase Half-Wave Rectifier
- Uncontrolled Single-Phase Full-Wave Rectifier
- Controlled Single-Phase Full-Wave Rectifier
- Uncontrolled 3-Phase Full-Wave Rectifier
- Buck Converter
- Boost Converter
- Buck-Boost Converter
- Full-Bridge Inverter with R-L Load

The following equations and figures are relevant for knowledge area 2.B.2, Power Supplies and Converters.

Uncontrolled Single-Phase Half-Wave Rectifier

Handbook: Uncontrolled Single-Phase Half-Wave Rectifier

$$V_o = V_{\text{avg}} = \frac{1}{2\pi} \int_0^T V_m \sin \omega t \; d\omega t = \frac{V_m}{\pi}$$

A single-phase half-wave rectifier passes one half of the AC input voltage for conversion to DC output and blocks the other half of the AC input voltage. An uncontrolled rectifier uses a diode, giving the fixed DC output voltage. The *Handbook* equation calculates the output voltage, V_o, or the average DC voltage, V_{avg}. V_m is the peak voltage. The following figure is adapted from *Handbook* figure Uncontrolled Single-Phase Half-Wave Rectifier.

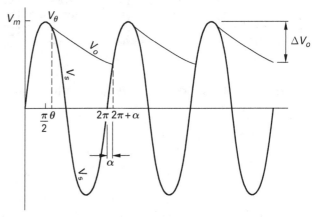

See also EPRM Fig. 34.13(a), which shows the circuit diagram of a half-wave rectifier and its output.

The *Handbook* includes an additional equation for a small variation in the direct current voltage, ΔV_o, after rectification of the AC voltage. V_m is the peak voltage, and V_o is the output voltage.

$$\Delta V_o \approx V_m \left(\frac{2\pi}{\omega RC} \right) = \frac{V_m}{fRC}$$

Controlled Single-Phase Half-Wave Rectifier

Handbook: Controlled Single-Phase Half-Wave Rectifier

$$V_o = \frac{1}{2\pi} \int_{\alpha}^{\pi} V_m \sin \omega t \, d\omega t = \left(\frac{V_m}{2\pi}\right)(1 + \cos \alpha)$$

A controlled rectifier uses silicon-controlled rectifiers (SCRs) that can be fired at the desired time to determine the amount of conduction time in a given cycle.

α represents the firing angle or delay angle. Conduction occurs only once the firing occurs. The delay is from zero crossing of the signal to SCR conduction. V_o is the output voltage, and V_m is the peak voltage.

Uncontrolled Single-Phase Full-Wave Rectifier

Handbook: Uncontrolled Single-Phase Full-Wave Rectifier

$$v_o(t) = V_o + \sum_{n=2,4,\cdots}^{\infty} V_n \cos(n \, \omega_0 t + \pi)$$

$$V_o = \frac{2 V_m}{\pi}$$

$$V_n = \frac{2 V_m}{\pi}\left(\frac{1}{n-1} - \frac{1}{n+1}\right)$$

A full-wave rectifier uses both half cycles of the input AC voltage and converts it into DC voltage. The uncontrolled full-wave rectifier uses two or four diodes for rectification. This equation is a Fourier series representation of the voltage. It is a combination of the base voltage from the source and the even harmonics. The variable n in the equation represents the number of the even harmonics.

Controlled Single-Phase Full-Wave Rectifier

Handbook: Controlled Single-Phase Full-Wave Rectifier

$$V_o = \frac{1}{\pi} \int_{\alpha}^{\pi} V_m \sin \omega t \, d\omega t = \left(\frac{V_m}{\pi}\right)(1 + \cos \alpha)$$

This equation determines the output voltage of a controlled single-phase full-wave rectifier. V_o is the output voltage, and V_m is the peak voltage.

EPRM Fig. 34.13(b) shows the circuit diagram of a full-wave bridge rectifier and its output.

Uncontrolled 3-Phase Full-Wave Rectifier

Handbook: Uncontrolled 3-Phase Full-Wave Rectifier

$$V_0 = \frac{1}{\left(\frac{\pi}{3}\right)} \int_{\pi/3}^{2\pi/3} V_{m,LL} \sin \omega t \, d\omega t = \frac{3 V_{m,LL}}{\pi} = 0.955 \, V_{m,LL}$$

An uncontrolled three-phase full-wave rectifier converts three-phase AC voltage input to the DC voltage output. It uses six diodes in total, with two diodes used per phase. $V_{m,LL}$ is the peak voltage (line-to-line), and V_0 is the output voltage.

Buck Converter

Handbook: Buck Converter

$$D = \frac{t_{on}}{T}$$

$$V_o = V_s D$$

A buck converter is a DC-DC converter that bucks the input, resulting in a step down of the output voltage.

D is the duty cycle. The value varies from 0 to 1.

t is the time the converter is in the ON state during the commutation period T.

V_o is the output voltage, and V_s is the source voltage.

EPRM Fig. 27.15 shows the autotransformer connections. An autotransformer is also called a buck or boost transformer, depending on the connection.

Figure 27.15 *Autotransformer Connections*

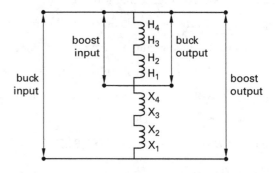

Boost Converter

Handbook: Boost Converter

$$V_o = \frac{V_s}{1 - D}$$

A boost converter boosts the input, which raises the output voltage level. V_o is the output voltage and V_s is the source voltage. D is the duty cycle.

Buck-Boost Converter

Handbook: Buck-Boost Converter

$$V_o = -V_s\left(\frac{D}{1-D}\right)$$

A buck-boost converter either lowers or boosts the voltage as required. The output voltage, V_o, depends on the duty cycle, D. V_s is the source voltage.

Full-Bridge Inverter with R-L Load: Instantaneous Current

Handbook: Full-Bridge Inverter with R-L Load

$$i_o(t) = \begin{cases} \dfrac{V_{DC}}{R} + \left(I_{\min} - \dfrac{V_{DC}}{R}\right)e^{-t/\tau} & \text{[for } 0 < t < T/2] \\ \dfrac{-V_{DC}}{R} + \left(I_{\max} + \dfrac{V_{DC}}{R}\right)e^{(t-T/2)/\tau} & \text{[for } T/2 < t < T] \end{cases}$$

The *Handbook* equation determines the current waveform for the R-L load.

A complementary MOSFET (CMOS) device used an inverter is illustrated in EPRM Fig. 37.19.

Figure 37.19 CMOS Inverter

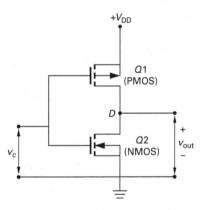

(a) digital switch

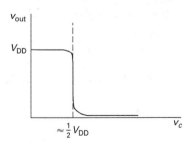

(b) characteristics

Additional circuitry would be required, but this device provides a fully electronic version of the switching that could be used in a full bridge inverter.

Full-Bridge Inverter with R-L Load: Max and RMS Current

Handbook: Full-Bridge Inverter with R-L Load

$$I_{\max} = -I_{\min} = \left(\frac{V_{DC}}{R}\right)\left(\frac{1 - e^{-T/2\tau}}{1 + e^{-T/2\tau}}\right)$$

$$I_{rms} = \sqrt{\frac{1}{T}\int_0^T i^2(t)\,dt} = \sqrt{\frac{2}{T}\int_0^{T/2}\left(\frac{V_{DC}}{R} + \left(I_{\min} - \frac{V_{DC}}{R}\right)e^{-t/\tau}\right)^2 dt}$$

The *Handbook* equations determine the maximum/minimum current and the rms current for a full-bridge inverter with an R-L load.

3. RELAYS, SWITCHES, AND LADDER LOGIC

Relays, switches, and ladder logic are not specifically covered in the *NCEES Handbook*. Relays are only mentioned in single-line diagrams, and switches are mentioned throughout but not explained in detail. Ladder logic is not mentioned in the *Handbook* in spite of its widespread use in programmable logic arrays (PLA). Nevertheless, some of the more important principles are likely to be covered under some portion of the exam.

Knowledge Area Overview

Key concepts: These key concepts are important for answering exam questions in knowledge area 2.B.3, Relays, Switches, and Ladder Logic.

- relays, switches, and ladder logic
- types of switches
- the normal state of a switch
- ladder logic diagrams and symbols
- ladder logic in programmable logic controllers

PE Power Reference Manual (**EPRM**): Study these sections in EPRM that either relate directly to this knowledge area or provide background information.

- Section 26.4: Overcurrent Protection
- Chapter 31: Protection and Safety
- Section 37.28: Ladder Logic
- Section 37.29: Ladder Logic: Timers

NCEES Handbook: The *Handbook* does not include any sections for this knowledge area.

The following equations are relevant for knowledge area 2.B.3, Relays, Switches, and Ladder Logic.

Relays

EPRM: Sec. 31.8

In a protection system, relays monitor the input from the transducers and provide an output to operate breakers in the event of a fault.

A protection system is comprised of three main elements: transducers, relays, and breakers. Transducers sense the desired signal and convert it into a form usable by the relays. Current and voltage instrument transformers are examples of transducers. Breakers are designed to interrupt power and isolate sections of the power system.

Universal Relay Equation

EPRM: Sec. 31.11

$$\tau = \tau_{\text{magnetic}} - \tau_{\text{spring}}$$
$$= C_1 I^2 + C_2 V^2 + C_3 IV \sin(\theta - \phi) - \tau_{\text{spring}} \qquad \textit{31.1}$$

EPRM Eq. 31.1 is called the universal relay equation. This equation can be used with solid-state relays or programmed into computer relays. θ is the torque angle in an analog system during normal conditions and is related directly to the impedance angle. ϕ is the torque angle in an analog system at the referenced (potentially, the fault) condition, which is also related to the impedance angle. By manipulating the constants in either an analog or a digital protective system, a variety of relays can be realized. EPRM Fig 31.16 shows the relay types that can occur.

Ladder Logic

EPRM: Sec. 37.28–Sec. 37.29

Ladder logic is a rule-based programming language used primarily in programmable logic controllers (PLCs).

Ladder logic symbols are used in diagrams to document connections between devices such as

- contactors
- relays or coils
- lamps or alarms

See EPRM Sec. 37.28 and Sec. 37.29 for more in-depth information on ladder logic symbols and timers.

4. VARIABLE SPEED DRIVES

A variable speed drive (VSD) is an electrical device whose speed varies across a considerable range as a function of the load. In general, a VSD refers to either AC or DC drives. A variable frequency drive (VFD) refers to AC drives only. VFD devices change the frequency while VSD devices change the voltage, generally.

Knowledge Area Overview

Key concepts: These key concepts are important for answering exam questions in knowledge area 2.B.4, Variable Speed Drives.

- advantages of using VSDs and VFDs
- speed in an AC motor using VSDs and VFDs
- energy savings due to use of VSDs and VFDs

PE Power Reference Manual **(EPRM):** Study these sections in EPRM that either relate directly to this knowledge area or provide background information.

Section 33.33: Variable Frequency Drive (VFD)

NCEES Handbook: To prepare for this knowledge area, familiarize yourself with this section in the *Handbook.*

- Variable-Speed Drives
- Induction Machines: Synchronous Speed

The following equations and figures are relevant for knowledge area 2.B.4, Variable-Speed Drives.

Variable Frequency Drives (VFD)

Handbook: Variable-Speed Drives

EPRM: Sec. 33.33

The *Handbook* contains a single paragraph on variable speed drives (VSD) and no equations. A more in-depth explanation of the basics of variable speed drives is found in EPRM Sec. 33.33. Use EPRM as the starting point for study of the important characteristics.

The VFD changes the frequency in order to change the speed of the rotational field, as shown by EPRM Eq. 33.71.

$$np = 120f \qquad \textit{33.71}$$

The number of poles, p, is generally fixed. Changing the frequency, f, changes the speed of the rotational field, n, which varies the overall speed of the motor.

Variable Frequency Drive (VFD) Functional Schematic

EPRM: Sec. 33.33

The variable frequency drive (VFD) is an AC drive. EPRM Fig. 33.25 shows a block diagram of a VFD. In Fig. 33.26, the VFD shown is a six-pulse device.

Figure 33.25 *VFD Block Diagram*

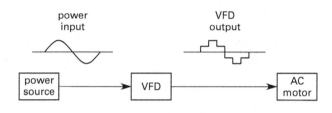

Figure 33.26 *VFD Functional Schematic*

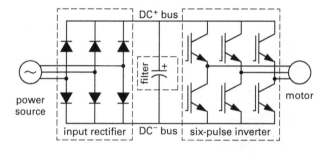

The insulated gate bipolar transistors (IGBTs) in the inverter section use pulse-width modulation (PWM) to create an easily modified square wave whose frequency and overall value can be changed. The modifications produce a sine wave.

5. DEFINITIONS

anode: The (positive) electrode from which conventional current enters a device. Note that this is conventional current, that is, positive to negative current flow. Anions (negative charges in solution) are attracted to the anode.

boost converter: A DC to DC power converter that raises the output voltage.

buck converter: A DC to DC power converter that lowers the output voltage.

buck-boost converter: A DC to DC power converter that either lowers or raises the output voltage depending on usage.

cathode: The (negative) electrode from which conventional current leaves a device. Use the memory aid "cathode current departs". Note that this is conventional current, that is, positive to negative current flow. Cations (positive charges in solution) are attracted to the cathode.

converter: An electrical circuit that transforms alternating current (AC) input into direct current (DC) output.

cutoff voltage: The voltage at which a battery was designed to be discharged to. Discharging further than the cutoff voltage could destroy the battery.

C-rate: The measure of the rate a battery is discharged relative to its maximum capacity.

dry cell battery: A primary battery that contains no free or liquid electrolyte.

electrochemical equivalent: The mass of a substance deposited to an electrode when 1 A passes for 1 second, that is, the mass that is transported by 1 C (one coulomb) of charge.

E-rate: The discharge power, in watts, at which a battery will discharge completely in one hour.

inverter: An electrical circuit that transforms direct current (DC) input into alternating current (AC) output.

ladder logic: A rule-based programming language used primarily in programmable logic controllers (PLCs).

primary battery: A type of battery that is designed to be used once until discharged and then discarded.

rectifier: An AC to DC power converter that works by reversing the AC current to DC current by only allowing it to flow in one direction.

relay: An electrically operated switch that opens and closes a circuit by receiving electrical signals from an outside source.

secondary battery: A type of battery that is designed to be discharged and recharged multiple times.

switch: A device for making and breaking the connection in an electric circuit. The switch can be a physical switch or an electrical one, such as a transistor operating in the switching mode.

variable frequency drive: Controls the speed of an AC motor by modulating the input frequency and voltage.

variable speed drive: Controls the speed and torque of a motor by changing a fixed input voltage and frequency to a variable voltage and frequency.

6. NOMENCLATURE

C	capacity	Ah/g, g/Ah, Ah/cm^3
C	charge	C
C	capacitance	F
Cp	Peukert's capacity	Ah
D	duty cycle	%
E	energy	J
E	voltage	V
EC	electrochemical equivalent	Ah/g, g/Ah, Ah/cm^3
f	frequency	Hz
H	discharge period	h
i	variable current	A
1	DC or rms current	A
k	Peukert's constant	–
M	multiple or fraction	–
n	number	–
n	number of the nth even harmonic	2, 4, 6...
n	speed	rpm
p	number of poles	–
P	power	watts
Q	charge	C
R	resistance	Ω
t	time	S
T	period	s
V	voltage	V
Z	impedance	Ω

Subscripts

0	initial, standard
a	anode
ave	average
avg	average
c	cathode
LL	line to line
m	max
max	maximum
min	minimum
n	nth voltage
n	time hours
o	output
p	Peukert
rms	root mean square
s	source
T	period

Symbols

α	firing angle	radians
θ	torque angle at normal condition	radians
τ	time constant	s
τ	torque	N·m
ϕ	torque angle at referenced condition	radian
ω	angular frequency	radian/s

6 Induction and Synchronous Machines

Exam specification 3.A, Induction and Synchronous Machines, makes up between 9% and 14% of the PE Electrical Power exam (between 7 and 11 questions out of 80).

The organization of this chapter follows the order of knowledge areas given by the NCEES for this exam specification. Each knowledge area is covered in the following numbered sections.

Content in blue refers to the *NCEES Handbook*.

Content in red is additional essential information.

1. GENERATOR/MOTOR APPLICATIONS

Generators and motors are the building blocks of many complicated systems. The theories behind the operation of generators and motors are similar. For generator action, a magnetic field, a conductor, and relative motion between the two is needed. For motor action, a current-carrying conductor in a magnetic field, which then results in motion between the two, is needed. The right-hand rule aids in determining the direction of the magnetic field and can be used for both generators and motors.

Knowledge Area Overview

Key concepts: These key concepts are important for answering exam questions in knowledge area 3.A.1, Generator/Motor Applications.

- systems with parallel generators and transformers
- calculation of slip
- system frequency
- generator droop
- two types of induction machines and how they work
- types of AC induction motors and classifications
- real and reactive powers will divide on parallel generators

- motor parameters and power transfer through induction motors
- maximum load that can be supplied to parallel transformers

PE Power Reference Manual (EPRM): Study these sections in EPRM that either relate directly to this knowledge area or provide background information.

- Section 7.21: Magnetic Poles
- Section 7.31: Speed and Direction of Charge Carriers
- Section 13.2: Magnetic Fields
- Section 23.6: Alternating Current Generators
- Section 23.7: Parallel Operation
- Section 23.8: Direct Current Generators
- Section 24.5: Generation of Three-Phase Potential
- Section 29.3: Three-Phase Connections
- Section 32.12: Torque and Power
- Section 33.8: Production of AC Potential
- Section 33.16: Induction Motors
- Section 33.17: Induction Motor Equivalent Circuit

NCEES Handbook: To prepare for this knowledge area, familiarize yourself with these sections in the *Handbook*.

- Synchronous Machines: Nomenclature
- Synchronous Machines: Power, Torque, and Speed Relationships
- Synchronous Machines: Synchronous Speed
- Governor Control for Synchronous Generators
- Induction Machines: Nomenclature
- Induction Machines: Power, Torque, and Speed Relationships
- Induction Machines: Synchronous Speed
- Percent Slip in Induction Machines
- Speed Control for Induction Machines
- Torque-Speed Characteristics with Varying Rotor Resistance

The following equations apply to knowledge area 3.A.1, Generator/Motor Applications.

Power, Torque, and Speed Relationships

Handbook: Synchronous Machines: Power, Torque, and Speed Relationships; Induction Machines: Power, Torque, and Speed Relationships

$$P_{kW} = T_{N \cdot m}\left(\frac{n_{rpm}}{9549}\right)$$

EPRM: Sec. 32.12

$$T_{N \cdot m} = \frac{9549 P_{kW}}{n_{rev/min}} = \frac{1000 P_{kW}}{\omega_{mech}} \qquad \textit{32.16}$$

The *Handbook* and EPRM equations are equivalent. Note the switched positions of the power and torque terms.

9549 is a conversion factor that comes from dividing the product of 60 sec/min and 1000 kW/W by 2π. The power is in kilowatts.

Synchronous Speed

Handbook: Synchronous Machines: Synchronous Speed; Induction Machines: Synchronous Speed

$$n_s = \frac{120 f}{p}$$

EPRM: Sec. 33.8

$$n_s = \frac{120 f}{p} = \frac{60 \omega_{mech}}{2\pi} = \frac{60 \omega}{\pi p} \quad \begin{bmatrix} \text{synchronous} \\ \text{speed} \end{bmatrix} \quad \textit{33.20}$$

The variable p is the number of poles. The more poles a motor has, the slower the motor's speed. EPRM Eq. 33.20 includes alternate forms that include the angular frequency, ω_{mech}.

The frequency, f, is

$$f = \frac{1}{T} = \frac{\omega}{2\pi} = \frac{p n_s}{120} \qquad \textit{33.21}$$

Speed Droop in Synchronous Generators

Handbook: Governor Control for Synchronous Generators

EPRM: Sec. 32.16

$$\text{speed droop\%} = \frac{n_{nl} - n_{fl}}{n_{fl}} \times 100\%$$

In EPRM Eq. 32.22, the equation is given in terms of speed regulation, SR, instead of speed droop percentage. However, the *Handbook* and EPRM equations are equivalent.

The frequency droop shown in EPRM Eq. 23.7 and Eq. 23.8 is directly related to the speed droop.

$$P = \frac{f_{sys} - f_{nl}}{f_{droop}} \qquad \textit{23.7}$$

$$P = \frac{f_{nl} - f_{sys}}{f_{droop}} \qquad \textit{23.8}$$

Equation 23.7 is normally in units of Hz/kW. The value is negative. The droop, though more properly a negative value, is often given as a positive value, in which case the formula used is Eq. 23.8.

Speed (frequency) controls the real load division among parallel generators.

Voltage controls the reactive load division among parallel generators.

Governor Control for Synchronous Generators

Handbook: Governor Control for Synchronous Generators

$$\Delta P = m_p (f_{nominal} - f_{actual})$$

m_p is the governor droop response in MW/Hz.

The real and reactive load sharing can be represented on the "house" diagrams shown. The name derives from their similarity to a house structure.

Figure 23.9 *Real Load Sharing*

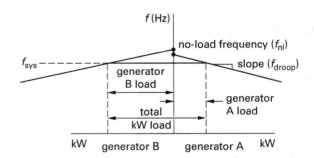

Figure 23.10 Reactive Load Sharing

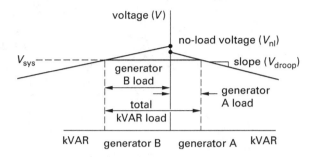

The system value is also called the nominal value.

The frequency or voltage controller actually changes the no-load setpoint, and the linear control line moves up or down, shifting the load between machines. The no-load frequency is

$$f_{nl} = Pf_{droop} + f_{sys}$$

Percent Slip in Induction Machines

Handbook: Percent Slip in Induction Machines

EPRM: Sec. 33.16

$$s\% = \frac{n_s - n}{n_s} \times 100\%$$

EPRM Eq. 33.40 is another form of this equation. s is a decimal fraction between zero and one, and ω is in radians per second.

$$s = \frac{\omega_{mech,s} - \omega_{mech}}{\omega_{mech,s}} \qquad 33.40$$

To generate a torque in a motor, a current-carrying conductor in a magnetic field is needed.

To generate the current in the rotor of the induction motor, there must be a conductor (the rotor), a magnetic field (the stators' magnetic field) and relative motion between the two—which can only occur if there is a difference in speed (called slip) between the two.

Speed Control for Induction Machines

Handbook: Induction Machines: Synchronous Speed

EPRM: Sec. 33.8

The *Handbook* does not provide any equations for speed control for induction machines. General information on the four common methods of speed control for induction machines is presented next.

The *Handbook* equation for synchronous speed illustrates how changing the number of poles, p, can result in a change of speed.

$$n_s = \frac{120f}{p}$$

The poles can be switched in and out by a variety of methods. Additionally, the *Handbook* equation illustrates how shifting the frequency using any available method also results in a shift of speed.

If the rotor is a wound-rotor type, the resistance of the rotor can be varied. This changes the current on the rotor and, since the torque is proportional to the current, the torque changes and thus the speed also changes.

Finally, a variable speed drive can be incorporated using scalar- and vector-control methods, both of which allow rated torque to be developed anywhere from zero to rated speed by controlling the strength of the magnetic field while maintaining the voltage constant.

2. EQUIVALENT CIRCUITS AND CHARACTERISTICS

The *NCEES Handbook* does not provide equations for equivalent circuits and characteristics, but the *Handbook* does provide diagrams of the synchronous machine equivalent circuit and induction machine equivalent circuits. The equivalent circuits of the synchronous and induction machines shown in the *Handbook* and the *PE Power Reference Manual* (EPRM) are equivalent. While terminology and symbology differ, the principles of operation remain the same.

Knowledge Area Overview

Key concepts: These key concepts are important for answering exam questions in knowledge area 3.A.2, Equivalent Circuits and Characteristics.

- equivalent circuit model for synchronous generators and motors

- induced voltage in DC machines

- speed, armature, and field control in DC machines

- different types of DC machine connections

- classifications of AC machines

- types of synchronous machines and how they work

PE Power Reference Manual (EPRM): Study these sections in EPRM that either relate directly to this knowledge area or provide background information.

- Section 26.7: Fault Analysis: Symmetrical

- Chapter 32: Rotating DC Machinery

- Chapter 33: Rotating AC Machinery

NCEES Handbook: To prepare for this knowledge area, familiarize yourself with these sections in the *Handbook*.

- Synchronous Machines: Nomenclature
- Synchronous Machine Equivalent Circuit
- Induction Machines: Nomenclature
- Induction Machine Equivalent Circuit

The following equations and figures are relevant for knowledge area 3.A.2, Equivalent Circuits and Characteristics.

Synchronous Motor Equivalent Circuit

Handbook: Synchronous Machine Equivalent Circuit

EPRM: Sec. 33.14–Sec. 33.15

For synchronous machines, the angle δ is the torque angle. It is related to the slip and is instrumental in stability analysis. The synchronous motor equivalent circuit is illustrated in EPRM Fig. 33.5.

Figure 33.5 Synchronous Motor Equivalent Circuit

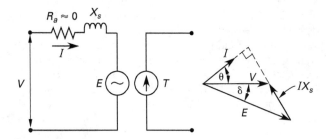

The synchronous generator equivalent circuit is illustrated in the following figure.

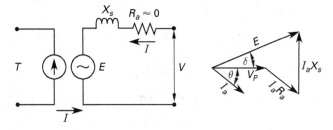

Note that there is a similarity with transformer equivalent circuits.

By changing the DC field current, the power factor that the machine presents to the system can be changed.

High field current is an overexcited state that produces VARs (Q). Lower field current is an underexcited state that absorbs VARs. The state in which the power is neither overexcited nor underexcited is known as *unity*.

The vector voltage relationship is defined by EPRM Eq. 33.33 and Eq. 33.34.

$$\mathbf{E} = \mathbf{V}_p + (R_a + jX_s)\mathbf{I}_a$$
$$\approx \mathbf{V}_p + jX_s\mathbf{I}_a \quad \text{[alternator]}$$

33.25

$$\mathbf{V}_p = \mathbf{E} + (R_a + jX_s)\mathbf{I}_a$$
$$\approx \mathbf{E} + jX_s\mathbf{I}_s \quad \text{[motor]}$$

33.26

Induction Motor Equivalent Circuit

Handbook: Induction Machine Equivalent Circuit

EPRM: Sec. 33.17

The traditional model of the equivalent circuit looks a lot like the transformer model. As shown in EPRM Fig. 33.13, a simplified model is used for most calculations.

Figure 33.13 Equivalent Circuits of an Induction Motor

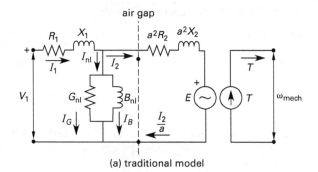

(a) traditional model

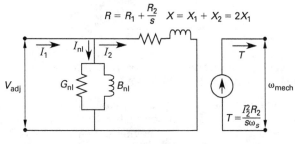

(b) simplified model ($a = 1$)

The variable a is the turns ratio. The air gap is analogous to the interface between the primary and secondary of a transformer.

R_1 is stator (winding) resistance. X_1 is stator (winding) reactance. G_{nl} is stator core loss ($G = 1/R$). B_{nl} is stator core susceptance ($1/X_{nl}$).

The dashed line represents the air gap across which energy is transferred to the rotor.

R_2 is rotor resistance. X_2 is rotor reactance.

The ratio of transformation, a, is 1.0 for a squirrel-cage motor.

The simplified model uses an adjusted voltage, V_{adj}. The relationship between the applied terminal voltage, V_1, and the adjusted voltage is shown in EPRM Eq. 33.41 and Eq. 33.42.

$$\mathbf{V}_{\text{adj}} = \mathbf{V}_1 - \mathbf{I}_{\text{nl}}(R_1 + jX_1) \qquad 33.41$$

$$V_{\text{adj}} \approx V_1 - I_{\text{nl}}\sqrt{R_1^2 + X_1^2} \qquad 33.42$$

DC Machine Connections

EPRM: Sec. 32.18

DC machines, both generators and motors, can be connected in a number of ways. EPRM Fig. 32.17 shows the various types of DC machine connections.

Figure 32.17 DC Machine Connections

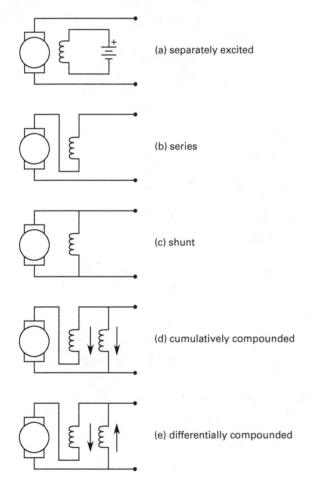

(a) separately excited

(b) series

(c) shunt

(d) cumulatively compounded

(e) differentially compounded

The first machine, Fig. 32.17(a), is separately excited. All the others, Fig. 32.17(b-e), are self excited. There are four types of self-excited DC machines: series, shunt, cumulatively compounded, and differentially compounded.

Series-Wired DC Motor Equivalent Circuit

EPRM: Sec. 32.19

The series-wired DC motor equivalent circuit is shown in EPRM Fig. 32.18. The equivalent circuit is very important to visualizing the field and the armature.

Figure 32.18 Series-Wired DC Motor Equivalent Circuit

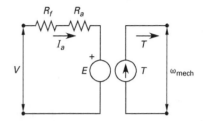

Note the differences between the series- and shunt-wired DC motor. The only components in the circuit are the field and armature resistances in a series.

I_a is positive for a motor and negative for a generator.

The torque is the mechanical output of the DC motor.

The speed, torque, and current are related.

$$\frac{T_1}{T_2} = \left(\frac{I_{a,1}}{I_{a,2}}\right)^2 \approx \frac{n_1}{n_2} \qquad 32.27$$

Shunt-Wired DC Machines

EPRM: Sec. 32.20

EPRM Fig 32.19 is a schematic of a DC machine controlled by the resistor in the shunt (parallel) field of the armature.

Figure 32.19 Equivalent Circuits

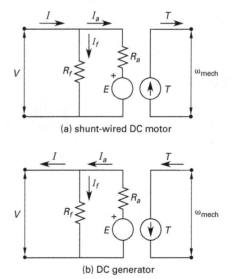

(a) shunt-wired DC motor

(b) DC generator

Note the differences between the shunt- and series-wired DC motor. The field resistance is in a different location. In the shunt-wired machine, the field current and armature current are not equal, as clearly shown in the equivalent circuit.

I_a is positive for a motor and negative for a generator.

The torque and current are proportional.

$$\frac{T_1}{T_2} = \frac{I_{a,1}}{I_{a,2}} \qquad 32.36$$

DC Motor Speed Control

EPRM: Sec. 32.25

The speed of a DC motor can be controlled by changing the armature conditions, field conditions, or both, as shown in EPRM Fig. 32.22.

Figure 32.22 *DC Motor Speed Control*

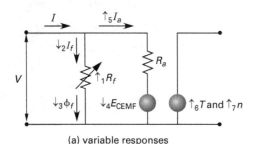

(a) variable responses

$\uparrow_1 R_f \rightarrow \downarrow_2 I_f \rightarrow \downarrow_3 \phi_f \rightarrow \downarrow_4 E_a$ or E_{CEMF}
$\rightarrow \uparrow_5 I_a \rightarrow \uparrow_6 T$ and $\uparrow_7 n$

(b) sequence of changes

Armature control techniques include (a) placing a variable resistance in series or parallel with the armature and (b) changing the voltage across the armature.

Two field control (field weakening) techniques include (a) changing the resistance of the field winding (series or shunt) and (b) changing the voltage across the field.

3. MOTOR STARTING

The *Handbook* does not provide any equations for motor starting. However, four starting methods commonly used with induction machines are provided in the *Handbook*. Each of these methods are reviewed next.

Knowledge Area Overview

Key concepts: These key concepts are important for answering exam questions in knowledge area 3.A.3, Motor Starting.

- how changing motor parameters can meet motor design objectives

- motor starting and speed control for AC motors

- motor starting and speed control for DC motors

- starting characteristics and applications for several types of motor starting

PE Power Reference Manual **(EPRM):** Study these sections in EPRM that either relate directly to this knowledge area or provide background information.

- Section 27.5: Three-Phase Transformer Configurations

- Section 29.11: IEEE Brown Book

- Chapter 32: Rotating DC Machinery

- Chapter 33: Rotating AC Machinery

NCEES Handbook: To prepare for this knowledge area, familiarize yourself with these sections and figures in the *Handbook*.

- Synchronous Machines: Nomenclature

- Synchronous Machines: Motor Starting

- Induction Machines: Nomenclature

- Induction Machines: Motor Starting

- 3-Phase Circuits

The following equations and figures are relevant for knowledge area 3.A.3, Motor Starting.

Across-the-Line Starting

Handbook: Induction Machines: Motor Starting

A motor can be started directly across the line. This occurs with full line voltage applied to the motor, resulting in a large power draw. The source voltage will drop. As long as this drop is acceptable, then this starting method is acceptable.

This method is simple and inexpensive.

The *Handbook* includes a figure showing a starting circuit that can be used with this method.

Rotor Resistance Starting

Handbook: Induction Machines: Motor Starting

If an induction motor is of the wound-rotor type, a high resistance may be inserted during starting to limit the current draw during starting.

Reduced Voltage Starting

Handbook: Induction Machines: Motor Starting

This method involves inserting an autotransformer between the motor and the source. The autotransformer, which varies the voltage from a low starting value to a higher value as the speed increases until the rated value is achieved, allows the current to be maintained at the desired levels.

The *Handbook* includes a figure showing the starting sequence that can be used with this method.

Wye Start/Delta Run Starting

Handbook: Induction Machines: Motor Starting

Since the line voltage of a wye circuit is less than that of a delta by a value of 1.71 (the square root of 3), the corresponding current is lower during starting if a wye circuit is used. When the speed is near the rated value, the circuitry switches to delta via contactors (or their electronic counterparts) and the motor comes up to full speed.

Three-Phase Line and Phase Relations

Handbook: 3-Phase Circuits

EPRM: Sec. 27.5

The following line and phase relationships apply to motor starting.

For delta connections,

$$V_L = V_P$$

$$I_L = \sqrt{3}\, I_P$$

For wye connections,

$$V_L = \sqrt{3}\, V_P = \sqrt{3}\, V_{\text{LN}}$$

$$I_L = I_P$$

EPRM Eq. 27.8–Eq. 27.11 use subscript ϕ to represent phase.

4. ELECTRICAL MACHINE THEORY

The basis of electrical machine theory is the production of potential voltages, the currents generated by such potential voltages, and the interactions of these currents. While generator designs vary in scope, they are fairly consistent in theory. Motors vary widely depending upon the principle used, from synchronous machines that lock fields, to induction machines where one field induces another, to reluctance machines that drag the rotor along with the stator's magnetic lines of flux.

Knowledge Area Overview

Key concepts: These key concepts are important for answering exam questions in knowledge area 3.A.4, Electrical Machine Theory.

- Faraday's law and the voltage induced in electrical machines
- types of power losses
- per-phase voltage, torque, and power
- relationship between torque and power
- analysis of parameters in the production of AC potential
- total developed torque
- voltage induced by moving a conductor through a magnetic field
- revolving and stationary field machines

PE Power Reference Manual (**EPRM**): Study these sections in EPRM that either relate directly to this knowledge area or provide background information.

- Section 7.21: Magnetic Poles
- Section 7.31: Speed and Direction of Charge Carriers
- Section 13.2: Magnetic Fields
- Section 29.3: Three-Phase Connections
- Chapter 32: Rotating DC Machinery
- Chapter 33: Rotating AC Machinery

NCEES Handbook: To prepare for this knowledge area, familiarize yourself with these sections in the *Handbook*.

- Synchronous Machines: Nomenclature
- Synchronous Machine Theory
- Induction Machines: Nomenclature
- Induction Machine Theory
- 3-Phase Induction Motor Power Flow

The following equations and figure are relevant for knowledge area 3.A.4, Electrical Machine Theory.

Synchronous Machine: Direct-Axis Synchronous Resistance

Handbook: Synchronous Machine Theory

EPRM: Sec. 33.14

$$X_{\text{ds}} = \frac{V_{\text{oc}}}{I_{\text{sc}}}$$

The direct axis synchronous reactance, X_{ds}, is determined using open-circuit and short-circuit tests.

The *Handbook* equation and EPRM Eq. 33.30 are the same.

Short-Circuit Ratio

Handbook: Synchronous Machine Theory

EPRM: Sec. 33.14

$$\text{SCR} = \frac{1}{X_{ds(pu)}}$$

The unsaturated short-circuit ratio (SCR) is the inverse of direct axis synchronous reactance and is given in per-unit terms. The SCR has applications in fault analysis.

The *Handbook* equation and EPRM Eq. 33.31 are the same.

Synchronous Machine Power: Cylindrical Rotor

Handbook: Synchronous Machine Theory

EPRM: Sec. 33.14

$$P_e = \frac{3E_0E}{X_s} \sin\delta \quad \text{[cylindrical rotor]}$$

P_e represents the steady-state power for a cylindrical rotor. The torque can be obtained by dividing the steady-state power by the synchronous speed. The *Handbook* equation and EPRM Eq. 33.32 are the same.

$$T_e = \frac{3E_0E}{X_s n_s} \sin\delta \quad \text{[cylindrical rotor]}$$

Synchronous Machine Power: Salient Pole Rotor

Handbook: Synchronous Machine Theory

EPRM: Sec. 33.14

$$T_e = \frac{3E_0E}{X_s n_s} \sin\delta$$

This equation represents the steady-state power for a salient pole rotor. It is more complex than the equation for a cylindrical rotor and involves two reactances, the direct and the quadrature reactance. The *Handbook* equation and EPRM Eq. 33.33 are the same.

The torque is obtained by dividing the steady-state power by the slip.

$$T_e = \frac{3E_0E}{X_s n_s} \sin\delta + \frac{3E^2(X_d - X_q)}{2X_d X_q n_s} \sin 2\delta \qquad \textit{33.35}$$

Synchronizing Power

Handbook: Synchronous Machine Theory

EPRM: Sec. 33.14

$$P_{\text{sync}} = \frac{3E_0E}{X_s} \cos\delta \quad \text{[cylindrical rotor]}$$

$$P_{\text{sync}} = \frac{3E_0E}{X_d} \cos\delta$$
$$+ \frac{3E^2(X_d - X_q)}{X_d X_q} \cos 2\delta \quad \begin{bmatrix} \text{salient pole} \\ \text{rotor} \end{bmatrix}$$

The synchronizing power is the varying of the power on torque angle (or load angle). It is also called the stiffness of coupling, stability, or rigidity factor. Other definitions also apply. See EPRM for a more in-depth explanation of the relationship to torque.

The *Handbook* equations and EPRM Eq. 33.36 and Eq. 33.37 are the same.

Induction Machine: Rotor Air Gap Power

Handbook: Induction Machine Theory

EPRM: Sec. 33.17

$$P_r = 3I_2^2 \frac{R_2}{s}$$

This is the rotor air gap power.

The subscript 2 represents the rotor values referred to the stator, much as in a transformer.

The *Handbook* equation and EPRM Eq. 33.46 are the same.

Mechanical Power Output

Handbook: Induction Machine Theory

EPRM: Sec. 33.17

$$P_m = (1 - s)P_r$$

Rotor copper losses are the slip multiplied by the rotor power, sP_r, hence this formula for the mechanical power output.

The *Handbook* equation and EPRM Eq. 33.47 are the same.

Mechanical Torque

Handbook: Induction Machine Theory

EPRM: Sec. 33.17

$$T_m = \frac{P_m}{(1-s)\omega_s} = \frac{pP_r}{4\pi f}$$

The *Handbook* provides this equation as the gross mechanical torque. The term p is the number of poles.

Both windage and friction are neglected in this equation. The *Handbook* equation and EPRM Eq. 33.48 are the same.

All the losses are shown in EPRM Fig. 33.15.

Figure 33.15 *Induction Motor Power Transfer*

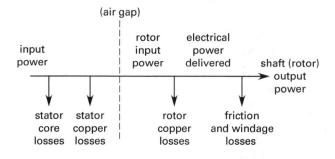

Electromagnetic Torque

Handbook: Induction Machine Theory

$$T_e = \left(\frac{3}{\omega_s}\right)\left(\frac{E^2\left(\dfrac{R_2}{s}\right)}{\left(R_1 + \dfrac{R_2}{s}\right)^2 + (X_1 + X_2)^2}\right)$$

This equation relates all the values to the rotor. The assumption is built in that the stator core reactance is much larger than the reactance of the stator windings—which allows for simplification.

In EPRM Fig. 33.15, the electromagnetic torque is the electric power delivered.

Core losses are independent of the load. In DC and synchronous AC machines, core losses occur in the armature iron. In induction machines, core losses occur in the stator iron.

Copper losses are real power losses due to wire and winding resistance.

Mechanical losses are due to brush and bearing friction and windage.

Condition for Maximum Torque

Handbook: Induction Machine Theory

$$\frac{R_2}{s_{\max T}} = \sqrt{R_1^2 + (X_1 + X_2)^2}$$

This *Handbook* equation represents the conditions for which maximum torque occurs in any given machine.

The term $s_{\max T}$ is the slip at which the breakdown torque occurs.

Three-Phase Induction Motor Power Flow

Handbook: 3-Phase Induction Motor Power Flow

$$P_{\text{in}} = \sqrt{3}\ V_T I_L \cos\theta$$

EPRM: Sec. 29.3

$$P = \sqrt{3}\ V_l I_l \cos\theta = \sqrt{3}\ VI\,\text{pf} \qquad \text{29.4}$$

The terminal voltage and the line voltage are generally considered the same. The designation of the voltage as "terminal" in the *Handbook* equation is meant to remove any consideration of line losses into the calculation.

5. DEFINITIONS

armature: The part of an electric machine in which the electromotive force is induced.

copper losses: Real power losses due to wire and winding resistance.

core losses: Constant losses that are independent of the load, including hysteresis and eddy current losses.

droop control: A regulatory mode whereby the load of a machine (due to its no-load setpoint and predetermined slope characteristics) determines the frequency or voltage value of the system.

induction machine: An AC electric motor in which the electric current in the rotor needed to produce torque is obtained by electromagnetic induction from the magnetic field of the stator winding.

mechanical losses: Losses due to brush and bearing friction and windage. Also known as *rotational losses*.

reactive load: A load carried by an AC system in which the current and voltage are out of phase and which is measured in volt-amperes or kilovolt-amperes.

rotor: The rotating part of a system.

slip: The difference between the synchronous speed and the actual speed of the rotor.

stator: The stationary part of a rotating system.

synchronous machine: An electrical machine whose rotating speed is proportional to the frequency of the alternating current supply and independent of the load.

synchronous speed: The speed of rotation of the magnetic field in a synchronous machine.

turns ratio: The ratio of the primary windings of a transformer to its secondary windings.

unity: The state in a synchronous motor or generator in which power is neither overexcited nor underexcited.

6. NOMENCLATURE

a	turns ratio	–
B	magnetic flux density	T
B	susceptance	S
E	electromotive force	V
f	frequency	Hz
G	conductance	S
I	current	A
m	droop (power)	MW/Hz
n	speed	rpm
p	number of poles	–
P	power	W, kW
pf	power factor	–
R	resistance	ohms
s	slip	%
SCR	short-circuit ratio	–
T	torque	N·m
V	voltage	V
X	reactance	Ω
Z	impedance	Ω

Symbols

δ	torque angle	radians
θ	angle	°
ω	angular frequency	radians/s

Subscripts

0	original
1	equivalent stator
2	equivalent rotor
a	armature, phase a
adj	adjusted
B	susceptance
d	direct axis
ds	direct axis synchronous
e	electromagnetic
f	field
fl	full load
G	conductance
l	line
L	line
LN	line-to-neutral
m	mechanical
mech	mechanical
N	neutral
nl	no load
oc	open circuit
p	power
P	phase
pu	per unit
q	quadrature axis
r	rotor
rpm	revolutions per minute
s	secondary, slip, or synchronous
sc	short circuit
sys	system
T	terminal
T	torque

7 Electric Power Devices

Exam specification 3.B, Electric Power Devices, makes up between 9% and 14% of the PE Electrical Power exam (between 7 and 11 questions out of 80).

The organization of this chapter follows the order of knowledge areas given by the NCEES for this exam specification. Each knowledge area is covered in the following numbered sections.

Content in blue refers to the *NCEES Handbook*.

Content in red is additional essential information.

1. TRANSFORMERS

A transformer, which is a magnetically coupled circuit, is used to change voltages, match impedances, and isolate circuits. A transformer can also raise or lower the values of capacitors, inductors, and resistors. Applications of transformers include the efficient transmission of electrical energy at high voltages over greater distances since transformers can lower voltages to safe values for industrial, commercial, and household use.

Knowledge Area Overview

Key concepts: These key concepts are important for answering exam questions in knowledge area 3.B.1, Transformers.

- buck and boost autotransformers
- connections of three-phase transformers
- division of real and reactive powers on parallel generators
- ideal and real transformers
- maximum capacity
- maximum load supplied to parallel transformers
- operation of open-delta transformers
- phase shift and paralleling conditions
- properties of three-phase transformers

- real transformer parameters
- secondary voltage and current
- step-down and step-up transformers
- total impedance
- turns ratio and its relationship to current, voltage, and impedance
- transformer's percentage voltage regulation
- voltages and currents through various transformer connections

PE Power Reference Manual (**EPRM**): Study these sections in EPRM that either relate directly to this knowledge area or provide background information.

- Section 17.19: Real Power and the Power Factor
- Chapter 18: Transformers
- Chapter 27: Power Transformers

NCEES Handbook: To prepare for this knowledge area, familiarize yourself with these sections in the *Handbook*.

- Single-Phase Transformer Equivalent Circuits
- Ideal Transformer
- Transformer's Exact Equivalent Circuit
- Transformer's Approximate Equivalent Circuit
- Transformer Losses
- Transformer's Percentage Voltage Regulation
- Transformer's Efficiency
- Condition for Maximum Efficiency
- Single-Phase Transformers in Parallel
- Autotransformers

The following equations and figure are relevant for knowledge area 3.B.1, Transformers.

Electromotive Force (emf) Induced in the Primary Winding

Handbook: Single-Phase Transformer Equivalent Circuits

$$E_1 = 4.44 N_1 f \phi_{\max}$$

$$E_2 = 4.44 N_2 f \phi_{\max}$$

EPRM: Sec. 18.2

$$V_s(t) = \frac{-N_s(2\pi f)\Phi_m \cos\omega t}{\sqrt{2}} \qquad 18.3$$
$$= -4.44 N_s f \Phi_m \cos\omega t$$

The *Handbook* equations show the maximum value of the electromotive force (emf) induced in the primary or secondary winding. N_1 is the number of primary turns, N_2 is the number of secondary turns, f is the frequency, and $\phi_{\max}$ is the maximum magnetic flux.

EPRM Eq. 18.3 displays the time-varying effective voltage value. The negative sign in EPRM Eq. 18.3 is a result of Lenz's law, which demonstrates that the flux produced in the secondary winding opposes the flux in the primary winding from which the flux in the secondary winding is produced. In other words, the induced emf will cause a current flow such that the current would generate a magnetic field that would oppose the change that created it. Essentially, EPRM Eq. 18.3 is a statement of the conservation of energy.

Real Transformer

EPRM: Sec. 18.5

The properties of an ideal core are listed along with a model in EPRM Fig. 18.8. The exact and approximate equivalent circuit is shown in *Handbook* diagrams Transformer's Exact Equivalent Circuit and Transformer's Approximate Equivalent Circuit.

- **Exciting current, I_E.** The exciting current is the total current needed to produce the magnetic flux in the core.

- **Mutual reactance, X_m, and magnetizing current, I_m.** The magnetizing current is the current necessary to generate the flux in the core. It is 90° out of phase with the core current.

- **Core resistance, R_c, and core current, I_c.** The core resistance represents iron losses (see EPRM Sec. 18.6 and Sec. 18.7). The core current is the current necessary to overcome those losses. This current flows in the primary windings, but it is caused by resistive losses physically occurring in the core.

Figure 18.8 Real Transformer Equivalent Circuit

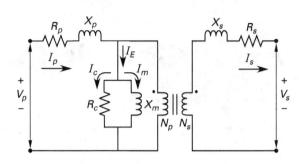

This model should be contrasted with the actual model, which exhibits the necessary currents to overcome the inevitable losses, however small. Real transformers may be 95% or higher in efficiency.

Turns Ratio

Handbook: Single-Phase Transformer Equivalent Circuits

EPRM: Sec. 18.3

$$a = \frac{N_1}{N_2}$$

The most common definition of turns ratio (or ratio of transformation) is shown in both the *Handbook* and the EPRM equation. But beware. There are texts and references that define the turns ratio as secondary to primary.

The terms shown in the *Handbook* equation are equivalent to those used in EPRM Eq. 18.10, where $N_1 = N_p$ and $N_2 = N_s$. For a step-down transformer, the turns ratio is greater than unity, and for a step-up transformer, the turns ratio is less than unity.

EPRM Eq. 18.20 shows the relationship of the turns ratio to the ratio of the primary to secondary leakage inductances.

$$a = \sqrt{\frac{L_p}{L_s}} \qquad 18.20$$

Ideal Transformer

Handbook: Single-Phase Transformer Equivalent Circuits

EPRM: Sec. 18.3

$$\frac{E_1}{E_2} = \frac{I_2}{I_1} = \frac{N_1}{N_2} = a$$

An ideal transformer has zero winding resistance, infinite core permeability, no leakage flux, and no core iron losses. The power transfer from the primary side equals the power to the secondary side. See *Handbook* diagram **Ideal Transformer**.

The terms shown in the *Handbook* equation are equivalent to those used in EPRM equations, where $I_1 = I_p$, $I_2 = I_s$, $N_1 = N_p$, and $N_2 = N_s$. The variables E_1 and E_2 are frequently used in the *Handbook* to indicate the induced voltage in the primary and secondary windings, respectively. They are equivalent to the variables V_p and V_s in EPRM.

For more information on ideal transformers, review EPRM Sec. 18.3.

Impedance

Handbook: Single-Phase Transformer Equivalent Circuits

$$R_{eq} = R_1 + a^2 R_2$$
$$X_{eq} = X_1 + a^2 X_2$$

EPRM: Sec. 18.4

$$Z_{ep} = Z_{ref} = Z_p + a^2 Z_s = \frac{V_p}{I_p} \qquad \textit{18.14}$$

The *Handbook* equation shows the impedance separated into the resistive and reactive components, while the EPRM equations combine the resistive and reactive components. R_1 and R_2, are the respective resistances of the primary and secondary windings. X_1, and X_2 are the respective leakage reactances.

In EPRM Eq. 18.14, Z_{ep} and Z_{ref} indicate the effective primary and reflective (or referred) values of the impedance.

The square of the turns ratio is a result of the relationship shown in EPRM Eq. 18.15.

$$a = \sqrt{\frac{Z_p}{Z_s}} \qquad \textit{18.15}$$

The approximate equivalent circuit is shown in *Handbook* diagram **Transformer's Approximate Equivalent Circuit**.

Secondary Voltage Referred to the Primary Side

Handbook: Single-Phase Transformer Equivalent Circuits

$$V_2' = a V_2$$

V'_2 is the secondary voltage referred to the primary side.

Review EPRM Chap. 18 and Chap. 27 for more details.

Secondary Winding Current Referred to the Primary Side

Handbook: Single-Phase Transformer Equivalent Circuits

$$I_2' = \frac{I_2}{a}$$

I'_2 is the secondary current referred to the primary side.

Secondary Winding Resistance Referred to the Primary Side

Handbook: Single-Phase Transformer Equivalent Circuits

$$R_2' = a^2 R_2$$

The *Handbook* equation calculates the secondary resistance referred to the primary side.

Secondary Winding Leakage Reactance Referred to the Primary Side

Handbook: Single-Phase Transformer Equivalent Circuits

$$X_2' = a^2 X_2$$

EPRM: Sec. 18.5

$$X_p = \omega L_p = \frac{V_{X_p}}{I_p} \qquad \textit{18.18}$$
$$= \frac{4.44 f \Phi_p N_p}{I_p}$$

$$X_s = \omega L_s = \frac{V_{X_s}}{I_s} = \frac{4.44 f \Phi_s N_s}{I_s} \qquad \textit{18.19}$$

In the *Handbook* equation, X'_2 is the secondary leakage reactance referred to the primary side.

The leakage reactance represents a flux value unable to produce a usable voltage in the desired output (the secondary). The *Handbook* and the EPRM equations are thus of the same form as those for the voltage produced —though with current accounted for in the denominator.

Secondary Winding Impedance Referred to the Primary Side

Handbook: Single-Phase Transformer Equivalent Circuits

$$Z_2' = R_2' + jX_2'$$

$$Z_2' = a^2 Z_2$$

EPRM: Sec. 18.4

$$Z_{ep} = Z_{ref} = Z_p + a^2 Z_s = \frac{V_p}{I_p} \qquad 18.14$$

The *Handbook* equation calculates the secondary winding impedance referred to the primary. R_2' and X_2' are the secondary winding resistance and reactance, respectively, referred to the primary side.

In EPRM Eq. 18.14, Z_{ep} and Z_{ref} indicate the effective primary and reflective (or referred) values of the impedance.

All values can be referred in either direction to calculate the desired quantity.

Transformer Output Power

Handbook: Single-Phase Transformer Equivalent Circuits

$$P_{out} = P_2 = V_2' I_2' \cos \theta_2$$

This is the standard equation for power expressed in terms of the output (secondary) voltage and current referred to the primary side.

The power factor, $\cos \theta_2$, is that of the secondary side because that is the load.

Transformer Input Power

Handbook: Single-Phase Transformer Equivalent Circuits

$$P_{inp} = P_1 = V_1 I_1 \cos \theta_1$$

EPRM: Sec. 17.19

$$P = I V \cos \phi_{pf} \qquad 17.53$$
$$= I V \, pf$$

This is the standard equation for power expressed in terms of the input (primary) voltage and current.

The power factor, $\cos \theta_1$, is that of the primary side because that is the load.

The *Handbook* and EPRM equations are equivalent. EPRM Eq. 17.53 uses the symbol ϕ instead of θ for the power factor angle.

Primary Winding Copper Losses

Handbook: Transformer Losses

$$P_{Cu} = I_1^2 R_1$$

EPRM: Sec. 27.2

$$P_{Cu} = I^2 R = I_1^2 R_p + I_2^2 R_s \qquad 27.4$$

The *Handbook* equation calculates the copper losses due to the wire resistance of the primary winding.

EPRM Eq. 27.4 includes both the primary and secondary copper losses. Wherever copper wires exist, such losses will occur.

Secondary Winding Copper Losses

Handbook: Transformer Losses

$$I_2^2 R_2 = I_2'^2 R_2'$$

The *Handbook* equation calculates the copper losses due to the wire resistance of the secondary winding. I_2' is the secondary winding current referred to the primary side, and R_2' is the secondary winding resistance referred to the primary side.

EPRM Eq. 27.4 shows both the primary and secondary copper losses. Wherever copper wires exist, such losses will occur.

$$P_{Cu} = I^2 R = I_1^2 R_p + I_2^2 R_s \qquad 27.4$$

Total Copper Losses

Handbook: Transformer Losses

$$P_{Cu} = I_1^2 R_1 + I_2^2 R_2$$

EPRM: Sec. 27.2

$$P_{Cu} = I^2 R = I_1^2 R_p + I_2^2 R_s \qquad 27.4$$

The copper losses, P_{Cu}, are caused by the total wire resistance. The *Handbook* equation and the EPRM equation are equivalent.

Transformer Core Losses

Handbook: Transformer Losses

$$P_c = P_{e+h} = P_e + P_h$$

The core losses consist of eddy current losses and hysteresis losses. Both of these losses are constant and independent of the transformer load.

In EPRM Eq. 27.3, core power losses are estimated by ignoring the small primary resistance in the model, R_p. R_c is the reciprocal of the core conductance.

$$P_c = \frac{V_1^2}{R_c} \qquad \text{27.3}$$

Eddy Current Losses

Handbook: Transformer Losses

EPRM: Sec. 27.2

$$P_e = K_e f^2 B_m^2$$

In the *Handbook* equation, the eddy current power, P_e, is given in W/kg. f is frequency, K is the coupling coefficient, and B_m is the maximum flux density in Wb/m^2.

Eddy current losses depend upon the mass of iron in the core. These losses are constant and independent of the transformer load. EPRM Eq. 27.1 includes the mass, m, in kg so that the resulting power is in units of watts, W. Besides this difference, the *Handbook* and the EPRM equations are equivalent.

Hysteresis Losses

Handbook: Transformer Losses

EPRM: Sec. 27.2

$$P_h = K_h f B_m^n \quad [n = 1.5 \rightarrow 2.5]$$

Hysteresis losses, P_h, are given in W/kg. The hysteresis losses are calculated using maximum flux density, B_m (in Wb/m^2), the Steinmetz exponent, n, (which varies from 1.5 to 2.5), the frequency, f, and the coupling coefficient, K.

Hysteresis current losses actually depend upon the mass of iron in the core. These losses are constant and independent of the transformer load. EPRM Eq. 27.2 includes the mass, m, in kg so that the resulting power is in units of watts, W. Besides this difference, the *Handbook* and the EPRM equations are equivalent.

Percentage Voltage Regulation

Handbook: Transformer's Percentage Voltage Regulation

$$\%\text{VR} = \frac{|V_{2\text{-nl}}'| - |V_{2\text{-fl}}'|}{|V_{2\text{-fl}}'|} \times 100\%$$

$$\%\text{VR} = \frac{I_{\text{op}}}{I_{\text{ra}}}\left(\%R\cos\theta + \%X\sin\theta + \frac{(\%X\cos\theta - \%R\sin\theta)^2}{200} \right)$$

EPRM: Sec. 27.4

$$\text{VR} = \frac{V_{\text{nl}} - V_{\text{fl}}}{V_{\text{fl}}} \qquad \text{27.6}$$

$$\text{VR} = \frac{\dfrac{V_p}{a} - V_{s,\text{rated}}}{V_{s,\text{rated}}} \qquad \text{27.7}$$

The voltage regulation of a transformer is measured similarly to that of a distribution system. The first *Handbook* equation and EPRM Eq. 27.6 are equivalent.

Another method of measuring voltage regulation is using the rated quantities, hence the operating current, I_{op}, and rated current, I_{ra}, and "%" values for resistance and reactance in the second *Handbook* equation. EPRM Eq. 27.7 gives the same result using rated voltages.

Transformer's Efficiency

Handbook: Transformer's Efficiency

$$\eta = \frac{P_{\text{out}}}{P_{\text{out}} + (P_{\text{Cu}} + P_c)} \times 100\%$$

EPRM: Sec. 27.2

$$\eta = \frac{P_{\text{out}}}{P_{\text{in}}} = \frac{P_{\text{in}} - \sum P_{\text{losses}}}{P_{\text{in}}}$$

$$= \frac{P_{\text{out}}}{P_{\text{out}} + P_c + P_{\text{Cu}}} \qquad \text{27.5}$$

The transformer efficiency is the ratio of the output power to the input power. Efficiency, as with all machines, is calculated by dividing the desired power output (minus the losses) by the required power input to compensate for those losses.

Maximum Efficiency

Handbook: Condition for Maximum Efficiency

$$P_{\text{Cu}} = P_{\text{core}}$$

The transformer efficiency is maximized when the copper losses (or winding losses), P_{Cu}, equal the iron losses (or core losses), P_{core}; that is, the efficiency is greatest when the variable losses equal the constant losses. Core losses consist of eddy current losses and hysteresis losses.

Percentage Load at the Highest Efficiency

Handbook: Condition for Maximum Efficiency

$$\%\text{load} = \sqrt{\frac{P_{\text{core}}}{P_{\text{Cu}}}} \times 100\%$$

The *Handbook* equation gives the load percentage at the highest efficiency.

Information about copper losses, core losses, and the maximum efficiency is given in EPRM Sec. 27.2.

Single-Phase Transformers in Parallel

Handbook: Single-Phase Transformers in Parallel

$$\frac{I_1}{I_2} = \frac{\%Z_{T2}S_{T1}}{\%Z_{T1}S_{T2}}$$

$$\frac{S_{L1}}{S_{L2}} = \frac{\%Z_{T2}S_{T1}}{\%Z_{T1}S_{T2}}$$

Single-phase transformers are paralleled to minimize cost while increasing capacity. Care must be taken when doing so to ensure proper connections are made.

Though somewhat difficult to discern, what the *Handbook* equations show is that single-phase transformers in parallel will share the load based on the ratio of the reciprocals of their impedances. The highest impedance transformer will carry the smallest load. This occurs to ensure there is no overloading.

Related information on parallel transformers is found in EPRM Sec. 27.11.

To expand capacity, transformers may be connected for parallel operation if the following conditions between the two transformers are met: (1) similarly marked terminals must be connected, (2) ground connections must be compatible, and (3) the following values for the transformers must match.

- turns ratios
- primary voltages
- secondary voltages
- resistance and reactance (impedance values)

Voltage and Current Relationships of Autotransformers

Handbook: Autotransformers

$$\frac{V_L}{V_H} = \frac{N_C}{N_{\text{SE}} + N_C}$$

$$\frac{I_L}{I_H} = \frac{N_{\text{SE}} + N_C}{N_C}$$

An autotransformer consists of a single winding in two sections. One of the sections provides voltage by induction, just like a standard transformer. This section is the common portion of the winding (using the subscript C). The other section provides voltage by conduction only (using the subscript SE for series). Both sections of the winding provide voltage to the output. The winding is both the primary and the secondary winding.

The *Handbook* equations are merely a variation on the turns ratio. The subscripts L and H represent the low and high sides of the transformer, be it step-up, step-down, boost, or buck transformer.

Ratio of Input and Output Apparent Power to the Apparent Power Traveling Through the Windings

Handbook: Autotransformers

$$\frac{S_{\text{IO}}}{S_W} = \frac{N_{\text{SE}} + N_C}{N_{\text{SE}}}$$

EPRM: Sec. 27.14

$$\begin{aligned} S_{\text{at}} &= S_{\text{tw}}\left(1 + \frac{N_s}{N_p}\right) \\ &= S_{\text{tw}}\left(1 + \frac{N_2}{N_1}\right) \end{aligned}$$

27.44

An interesting aspect of a transformer is that when it is set up as an autotransformer, it has more capacity than originally intended.

The *Handbook* equation shows the ratio of the input and output apparent power, S_{IO}, to the apparent power flowing in the windings, S_W.

In EPRM Eq. 27.44, the apparent power of the autotransformer, S_{at}, is equal to the apparent power in a two-winding transformer, S_{tw}, multiplied by the standard turns ratio. Note that with the "1 +" term in front of the turns ratio, regardless of the setup, the capacity of the autotransformer is greater than that of the original two-winding transformer.

2. REACTORS

The terms *reactor* and *inductor* are sometimes used interchangeably. Specifically, a reactor is an electromechanical device (with inductors in it) used in power lines to limit short-circuit current to a safer value. An inductor consists of one or more windings with or without a core. Inductors are used for filtering, improving stability, or maintaining current.

Knowledge Area Overview

Key concepts: These key concepts are important for answering exam questions in knowledge area 3.B.2, Reactors.

- energy stored in an inductor

- function of inductors and factors affecting inductance

- inductances in parallel and in series

- relationship between inductance and reactive power

- voltage-current relationship for inductors

PE Power Reference Manual (**EPRM**): Study these sections in EPRM that either relate directly to this knowledge area or provide background information.

- Section 7.24: Permeability and Susceptibility

- Section 7.28: Inductance and Reciprocal Inductance

- Section 7.29: Inductors

- Section 7.32: Voltage and the Magnetic Circuit

- Section 17.15: Inductors

- Section 19.11: Inductance

NCEES Handbook: To prepare for this knowledge area, familiarize yourself with these sections in the *Handbook*.

- Reactors

- Inductor

The following equations and tables are relevant for knowledge area 3.B.2, Reactors.

Inductance

Handbook: Reactors

EPRM: Sec. 7.29

$$L = \frac{N^2 \mu A}{l} = \frac{N^2}{\mathcal{R}}$$

The *Handbook* equation gives the inductance, L, of a coil of N turns wound on a core with a cross-sectional area, A, permeability, μ, and flux with a mean path of l. $\mathcal{R}$ is the reluctance. See *Handbook* figure Inductor.

The *Handbook* equation and EPRM Eq. 7.67 are equivalent. Formulas for the inductance of other configurations are given in EPRM Table 7.8.

Reluctance

Handbook: Reactors

EPRM: Sec. 7.32

$$\mathcal{R} = \frac{l}{\mu A}$$

Reluctance, $\mathcal{R}$, in a magnetic circuit is analogous to the resistance in an electric circuit.

The *Handbook* equation is identical to EPRM Eq. 7.84. The reluctance is measured in A/Wb, which is equal to H^{-1}.

The analogous Ohm's law for magnetic circuits is shown in Eq. 7.85.

$$F_m = Hl = \phi \mathcal{R} \qquad \text{7.85}$$

Permeability of Free Space, μ_0

Handbook: Reactors

EPRM: Sec. 7.24

$$\mu = \mu_r \mu_0$$

Permeability is a measure of the ability of magnetic flux to penetrate a material. The *Handbook* equation gives permeability in terms of the permeability of free space, μ_0, and the relative permeability, μ_r, which depends upon the material used. The value of μ_0 is $4\pi \times 10^{-7}$ H/m.

The *Handbook* equation and EPRM Eq. 7.55 are equivalent.

Permeability can also be defined in terms of the magnetic flux density, B, and the magnetic field strength, H, as shown in Eq. 7.56.

$$\mu = \frac{B}{H} \qquad \text{7.56}$$

Permeability can also be expressed as in EPRM Eq. 7.57, where χ_m is a dimensionless quantity called the magnetic susceptibility.

$$\mu = \mu_0(1 + \chi_m) \qquad \text{7.57}$$

Faraday's Law

Handbook: Reactors

EPRM: Sec. 19.11

$$v_L(t) = L \frac{di_L}{dt}$$

$$i_L(t) = i_L(0) + \frac{1}{L} \int_0^t v_L(\tau)\, d\tau$$

The *Handbook* equation uses Faraday's law to show the voltage-current relationships for an inductor. The first *Handbook* equation and EPRM Eq. 19.16 are equivalent.

Consider the phrase "inductors oppose rates of change of current" as a memory aid. This accounts for two parts of the equation, L and di/dt, so the remaining part has to be voltage.

EPRM Table 19.1 shows the parameters of linear circuit elements.

EPRM Table 17.1 shows the characteristics of the passive elements in AC circuits.

Table 17.1 *Characteristics of Resistors, Capacitors, and Inductors*

	resistor	capacitor	inductor
value	R (Ω)	C (F)	L (H)
reactance, X	0	$\dfrac{-1}{\omega C}$	ωL
rectangular impedance, $\mathbf{Z}$	$R + j0$	$0 - \dfrac{j}{\omega C}$	$0 + j\omega L$
phasor impedance, $\mathbf{Z}$	$R\angle 0°$	$\dfrac{1}{\omega C}\angle -90°$	$\omega L\angle 90°$
phase	in-phase	leading	lagging
rectangular admittance, $\mathbf{Y}$	$\dfrac{1}{R} + j0$	$0 + j\omega C$	$0 - \dfrac{j}{\omega L}$
phasor admittance, $\mathbf{Y}$	$\dfrac{1}{R}\angle 0°$	$\omega C\angle 90°$	$\dfrac{1}{\omega L}\angle -90°$

Energy Stored in an Inductor

Handbook: Reactors

$$\text{energy} = \frac{Li_L^2}{2}$$

EPRM Sec. 19.11

$$U = \tfrac{1}{2}LI^2 = \tfrac{1}{2}\Psi I = \frac{1}{2}\left(\frac{\Psi^2}{L}\right) \qquad \textbf{19.20}$$

The *Handbook* equation determines the energy stored (in joules) in an inductor. EPRM Eq. 19.20 is the equivalent equation for the (average) energy stored in the magnetic field of an inductor.

A standard energy symbol is U. Sometimes E is used, but not as often in electrical engineering.

EPRM Table 19.1 shows the parameters of linear circuit elements.

EPRM Table 19.2 shows the time domain and the frequency domain behavior of linear circuit elements, along with their defining equations.

Relationship Between Inductance and Reactive Power

Handbook: Reactors

$$L_{\text{ph}} = \frac{V_{\text{ph}}^2}{2\pi f\left(\dfrac{\text{MVAR}}{\text{phase}}\right)}$$

L_{ph} is the inductance (in henrys) per phase, V_{ph} is the phase voltage (in kilovolts), and f is the frequency (in hertz). The *Handbook* equation is a rearrangement of one of the standard equations for reactive power. Be cautious when using the *Handbook* equation. Since units of MVAR are used for the reactive power, units of kilovolts need to be used for the phase voltage.

Similar equations can be derived from the reactive power equation given in EPRM Ex. 17.7.

$$Q = \frac{V_p^2}{X_L} = \frac{V_p^2}{\omega L} = \frac{V_p^2}{2\pi f L}$$

$$L = \frac{V_p^2}{2\pi f Q_p}$$

3. TESTING

Open-circuit and short-circuit tests are the two standardized tests performed. Parameters of transformers that are often tested include transformer rating, transformer efficiency, and equivalent circuit values of the real transformer.

Knowledge Area Overview

Key concepts: These key concepts are important for answering exam questions in knowledge area 3.B.3, Testing.

- transformer parameters from open-circuit test results

- transformer parameters from short-circuit test results

PE Power Reference Manual **(EPRM):** Study these sections in EPRM that either relate directly to this knowledge area or provide background information.

- Section 27.1: Theory

- Section 27.2: Transformer Rating

- Section 27.3: Transformer Capacity

NCEES Handbook: To prepare for this knowledge area, familiarize yourself with these sections and figures in the *Handbook*.

- **Open-Circuit Testing**

- **Single-Phase Transformer No-Load Equivalent Circuit**

- **No-Load Test Connection**

- **Short-Circuit Test**

- **Short-Circuit Test Connection**

- **Single-Phase Transformer Approximate Equivalent Circuit Under Short-Circuit Test**

The following equations and figures are relevant for knowledge area 3.B.3, Testing.

Open-Circuit Test

Handbook: Open-Circuit Testing

$$P_{\text{core}} = P_o$$

$$\cos \theta_o = \frac{P_o}{V_o I_o} \quad \text{[no-load power factor]}$$

$$I_c = I_o \cos \theta_o$$

$$I_m = I_o \sin \theta_o$$

$$R_c = \frac{V_o}{I_c}$$

$$X_m = \frac{V_o}{I_m}$$

EPRM: Sec. 27.7

$$Y_c = G_c + jB_c$$
$$= \frac{I_{1oc}}{V_{1oc}} \qquad 27.13$$

$$G_c = \frac{P_{oc}}{V_{1oc}^2} \qquad 27.14$$

$$B_c = \frac{1}{X_c}$$
$$= \frac{-1}{\omega L_c} \qquad 27.15$$
$$= -\sqrt{Y_c^2 - G_c^2}$$
$$= \frac{-\sqrt{I_{1oc}^2 \, V_{1oc}^2 - P_{oc}^2}}{V_{1oc}^2}$$

In the *Handbook* equations, P_o is the no-load power (core losses), and I_o is the no-load current measured during the test. V_o is the rated voltage applied. The core current, I_c, and the magnetizing current, I_m, are associated with the measured open-circuit test current.

These currents correlate with the EPRM model conductance (for the core current) and the susceptance (for the magnetizing) current. The susceptance is negative ($-1/\omega L$) for a lagging condition.

The equations for the core equivalent circuit parameters, R_c and X_m, are also given in the *Handbook*. The results of the open-circuit test for either model are equivalent; only the units are presented differently.

See *Handbook* figures **Single-Phase Transformer No-Load Equivalent Circuit** and **No-Load Test Connection**. EPRM Fig. 27.4 shows the open-circuit test model.

Figure 27.4 *Transformer Open-Circuit Test Model*

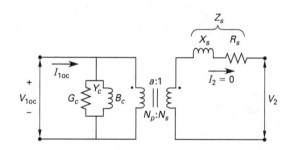

Short-Circuit Test

Handbook: Short-Circuit Test

EPRM: Sec. 27.8

$$P_{\text{Cu}} = P_{sc}$$

$$R_{eq} = \frac{P_{sc}}{I_{sc}^2}$$

$$|Z_{eq}| = \frac{V_{sc}}{I_{sc}}$$

$$X_{eq} = \sqrt{|Z_{eq}|^2 - R_{eq}^2}$$

P_{sc} is the power measured (copper losses). V_{sc} is the reduced voltage applied. Short-circuit tests are conducted at the rated current, where $I_{sc} = I_{ra}$. The equations in EPRM Sec. 27.8 are equivalent to the *Handbook* equations.

See *Handbook* figures Short-Circuit Test Connection and Single-Phase Transformer Approximate Equivalent Circuit Under Short-Circuit Test. EPRM Fig. 27.5 shows the short-circuit test model.

Figure 27.5 *Transformer Short-Circuit Test Model*

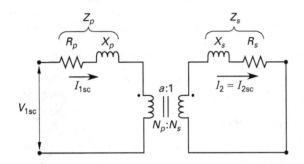

4. CAPACITORS

Capacitance is the property of conductors and dielectrics that permits the storage of charges. A capacitor charges and discharges the electric charge stored in it. The most common type of capacitor is a parallel plate capacitor.

Knowledge Area Overview

Key concepts: These key concepts are important for answering exam questions in knowledge area 3.B.4, Capacitors.

- capacitance

- energy storage in a capacitor

- function of a capacitor and the factors affecting capacitance

- relationship between capacitance and reactive power

- voltage-current relationship for a capacitor

PE Power Reference Manual (EPRM): Study these sections in EPRM that either relate directly to this knowledge area or provide background information.

- Section 7.9: Permittivity and Susceptibility

- Section 7.13: Capacitance and Elastance

- Section 7.14: Capacitors

- Section 19.10: Capacitance

- Section 33.25: Capacitor-Start Motors

- Section 33.27: Capacitor-Run Motors

NCEES Handbook: To prepare for this knowledge area, familiarize yourself with this section in the *Handbook*.

- Capacitors

The following equations and tables, are relevant for knowledge area 3.B.4, Capacitors.

Charge and Voltage Relationship for a Capacitor

Handbook: Capacitors

EPRM: Sec. 7.13

$$C = \frac{q_C(t)}{v_C(t)}$$

$$q_C(t) = Cv_C(t)$$

The *Handbook* equations and EPRM Eq. 7.22(a) and Eq. 7.22(b) are equivalent. The *Handbook* equations use instantaneous values, while the EPRM equations use total values. The capacitance, C, is in units of farads, F.

Capacitance of a Parallel Plate Capacitor with Plates Separated by an Insulator

Handbook: Capacitors

EPRM: Sec. 7.14

$$C = \frac{\epsilon A}{d}$$

The *Handbook* equation and EPRM Eq. 7.24 are equivalent. In the *Handbook* equation, the distance d represents the same value as r in the EPRM equation.

Just as inductance is dependent upon the permeability of the material, capacitance is dependent upon the permittivity, ϵ, of the material to allow the electric flux to penetrate.

Permittivity

Handbook: Capacitors

EPRM: Sec. 7.9

$$\epsilon = \epsilon_r \epsilon_0$$

The *Handbook* equation shows the permittivity, ϵ, in terms of relative permittivity, ϵ_r, and the permittivity of free space, ϵ_0. EPRM Eq. 7.11 is equivalent to the *Handbook* equation.

The value of ϵ_0 is 8.85×10^{-12} F/m. The relative permittivity for a vacuum is 1.0. The relative permittivity can also be expressed in terms of capacitance as given by EPRM Eq. 7.12.

$$\epsilon_r = \frac{C_{\text{with dielectric}}}{C_{\text{vacuum}}} \qquad 7.12$$

Current-Voltage Relationships for a Capacitor

Handbook: Capacitors

EPRM: Sec. 19.10

$$v_C(t) = v_C(0) + \frac{1}{C} \int_0^t i_C(\tau)\, d\tau$$

$$i_C(t) = C \frac{dv_C}{dt}$$

The *Handbook* equations show the current-voltage relationships for a capacitor. EPRM Eq. 19.9 is equivalent to the second *Handbook* equation.

Consider the phrase "capacitors oppose rates of change of voltage" as a memory aid. This accounts for two parts of the equation, C and dv/dt, so the remaining part has to be current.

Review EPRM Table 19.1 for the parameters of linear circuit elements.

Energy Stored in a Capacitor

Handbook: Capacitors

$$\text{energy} = \frac{Cv_C^2}{2} = \frac{q_C^2}{2C} = \frac{q_C v_C}{2}$$

EPRM: Sec. 19.10

$$U = \frac{1}{2}CV^2 = \frac{1}{2}QV$$
$$= \frac{1}{2}\left(\frac{Q^2}{C}\right) \qquad 19.13$$

The *Handbook* equation is equivalent to EPRM Eq. 19.13. U is the (average) energy stored in the electric field of a capacitor. The energy is expressed in joules.

Review EPRM Table 19.1 for the parameters of linear circuit elements.

Relationship Between Capacitance and Reactive Power

Handbook: Capacitors

$$C_{\text{ph}} = \frac{\left(\dfrac{\text{MVAR}}{\text{phase}}\right)}{2\pi f V_{\text{ph}}^2}$$

The *Handbook* equation is a rearrangement of one of the standard equations for reactive power. Be cautious when using the *Handbook* equation. Since units of MVAR are used for the reactive power, units of kV would need to be used for the phase voltage.

The following equation is derived using the reactive power equation given in EPRM Ex. 17.7 but uses capacitive inductance instead of inductive reactance.

$$Q = \frac{V_p^2}{X_C} = \frac{V_p^2}{\dfrac{1}{2\pi f C}} = V_p^2 2\pi f C$$

$$C = \frac{Q}{2\pi f V_p^2}$$

5. DEFINITIONS

autotransformer: A transformer with only one winding that is common to both the primary and secondary circuits associated with that winding.

copper loss: Resistive loss, also called I^2R loss, due to current flow through copper conductors.

coupling coefficient: A constant with a maximum value of 1 that is a measure of the flux linkage from one coil to another.

core loss: Constant loss that is independent of the load, including hysteresis and eddy current loss.

eddy current: Loops of electrical current induced in conductors (cores in the case of transformers) by the changing magnetic field. They are also called Foucault's currents. They are minimized by laminating the iron cores into small slices, thus minimizing the path lengths.

equivalent resistance or reactance: The amount of resistance or total reactance referred to the primary or the secondary side. The referred amount depends upon the turns ratio.

exciting current: The total current, I_E, required to produce the magnetic flux in the core. It is a combination of the core current, I_c, which is necessary to compensate for the iron losses, and the magnetizing current, I_m, which generates the flux in the core and is 90° out of phase with the core current.

hysteresis loss: The work done (hence a loss) by the magnetizing force against the internal friction of the molecules of the material.

iron loss: The combination of eddy current and hysteresis loss in a transformer.

reactive power: The product of the rms values of the current and voltage multiplied by the quadrature of the current.

reactor: Another name for an inductor.

real power: The average power consumed by the resistive elements of a circuit. Also known as true power.

turns ratio: The ratio of the primary windings of a transformer to its secondary windings.

6. NOMENCLATURE

a	turns ratio	
A	area	m^2
B	flux density	T
B	susceptance	S
B	magnetic flux density	Wb/m^2
C	capacitance	F
d	distance	m
E	induced voltage	V
E, V	voltage	V
f	frequency	Hz
F	force (magnetic)	V
G	conductance	S
H	magnetic field strength	A/m
i	instantaneous current	A
I	DC or rms current	A
K	coupling coefficient	–
L	inductance	H
m	mass	kg
N	number of turns	–
P	power	W (watts)
q	charge (instantaneous)	C
Q	reactive power	VAR
r	radius	m
R	resistance	Ω
$\mathcal{R}$	reluctance	A/Wb
S	apparent power	VA
t	time	s
U	energy	J
v	instantaneous voltage	V
V	DC or rms voltage	V
VR	voltage regulation	%
X	reactance	Ω
Y	admittance	S
Z	impedance	Ω

Symbols

ϵ	permittivity	F/m
η	efficiency	%
θ	angle	radians
μ	permeability	H/m
Ψ	flux	Wb
ϕ	flux	Wb
ω	angular frequency	radians/s

Subscripts

0	free space
1	primary
2	secondary
at	autotransformer
c	core
C	common
Cu	copper
e	eddy
ep	effective primary
eq	equivalent
fl	full load
h	hysteresis
H	high
inp	input
IO	ratio of input to output
L	inductor, low
m	magnetic, maximum
M	maximum
max	maximum

nl   no load

o no load

oc   open circuit

op   operating

p phase, primary

pf   power factor

ph   phase

r relative

ra   rated

ref   reflective

s secondary, source, synchronous

sc   short circuit

SE   series

T transformer

tw   two-winding

W windings

8 Power System Analysis

Exam specification 4.A, Power System Analysis, makes up between 13% and 19% of the PE Electrical Power exam (between 10 and 15 questions out of 80).

The organization of this chapter follows the order of knowledge areas given by the NCEES for this exam specification. Each knowledge area is covered in the following numbered sections.

Content in blue refers to the *NCEES Handbook*.

Content in red is additional essential information.

1. VOLTAGE DROP

Voltage drops are both advantageous and disadvantageous for proper operation of electrical equipment. They are useful for releasing energy, but they also result in heat loss. Limits are set by the circuit elements used, the length of conductors, the capabilities of the motors and equipment, and any standards invoked.

Knowledge Area Overview

Key concepts: These key concepts are important for answering exam questions in knowledge area 4.A.1, Voltage Drop.

- AC resistance

- DC resistance and resistivity

- percentage increase in resistivity due to temperature increase

- resistance in transmission lines

- skin depth and skin effect

PE Power Reference Manual **(EPRM):** Study these sections in EPRM that either relate directly to this knowledge area or provide background information.

- Chapter 16: DC Circuit Fundamentals

- Section 18.2: Magnetic Coupling

- Section 19.24: Kirchhoff's Voltage Law

- Section 24.2: Standard Notation Conventions

- Section 28.2: DC Resistance

- Chapter 44: National Electrical Code

- Section 44.7: Wiring and Protection: Feeders

NCEES Handbook: To prepare for this knowledge area, familiarize yourself with this section in the *Handbook*.

- Voltage Drop

The following equations are relevant for knowledge area 4.A.1, Voltage Drop.

Voltage Drop

Handbook: Voltage Drop

$$\mathrm{VD} = \frac{2LRI}{K \times 1000}$$

EPRM: Sec. 44.7

$$V_{\mathrm{drop}} = \frac{2lIR}{1000\ \mathrm{ft}} = IR_l 2l \qquad 44.4$$

The formula in the *Handbook* inserts a constant K, which is 1 for a single-phase or DC circuit. For a three-phase circuit, $K = 2/\sqrt{3}$. If the single-phase circuit is not purely resistive and the power factor is less than 1, K is adjusted.

The value of 1000 in the denominator of both equations is used because resistances for wire in the *National Electrical Code* are given in units of $\Omega/1000$ ft.

Line-to-Neutral Voltage Drop in a Three-Phase AC Circuit

Handbook: Voltage Drop

$$\mathrm{VD}_{\mathrm{ln}} \approx I\big(R(\mathrm{pf}) + X_L \sin\big(\cos^{-1}(\mathrm{pf})\big)\big)$$

The *Handbook* equation is a version of Ohm's law. The first expression is the voltage drop due to the real load. The second expression is the voltage drop due to the reactive load.

The arccos of the power factor angle gives the power angle, that is, the angle between the apparent power and the real power. Taking the sine of the result provides the reactive power magnitude.

2. VOLTAGE REGULATION

Voltage regulation, though simple in theory, has a wide impact. Proper circuit operation in a household computer requires the correct voltage range at the receptacle though the voltage is generated hundreds of miles from that location. Therefore, the voltage regulation along the entire path must be designed and controlled to the proper limits.

Knowledge Area Overview

Key concepts: These key concepts are important for answering exam questions in knowledge area 4.A.2., Voltage Regulation.

- neutral current
- positive and negative sequences
- total apparent power, real power, and power factor
- voltage regulation
- voltage regulation and speed regulation for electrical machines

PE Power Reference Manual **(EPRM):** Study these sections in EPRM that either relate directly to this knowledge area or provide background information.

- Chapter 16: DC Circuit Fundamentals
- Section 19.2: Ideal Independent Voltage Sources
- Section 27.4: Voltage Regulation
- Section 28.11: Transmission Line Representation

NCEES Handbook: To prepare for this knowledge area, familiarize yourself with this section in the *Handbook*.

- Voltage Regulation

The following equation is relevant for knowledge area 4. A.2, Voltage Regulation.

Voltage Regulation

Handbook: Voltage Regulation

EPRM: Sec. 16.8

$$\mathrm{VR} = \frac{V_{\mathrm{nl}} - V_{\mathrm{fl}}}{V_{\mathrm{fl}}} \times 100\% \qquad \textit{16.15}$$

Voltage regulation is the output of a voltage source. Independent sources provide voltage and current at their rated values, regardless of the circuit parameters. Dependent sources provide voltage and current levels based upon circuit parameters.

3. POWER FACTOR CORRECTION AND VOLTAGE SUPPORT

Power factor correction involves adding reactive load, usually negative, to balance the normally inductive reactive load on a system and bring the power factor closer to one. This is accomplished because the reactive power is not absorbed by the load and is instead passed back and forth from load to source. The source (and the conducting cables) must be able to handle this additional reactive power flow.

Knowledge Area Overview

Key concepts: These key concepts are important for answering exam questions in knowledge area 4.A.3, Power Factor Correction and Voltage Support.

- costs associated with power factor correction
- parts of a power factor correction system
- power factor
- required capacitance to obtain a desired power factor

PE Power Reference Manual **(EPRM):** Study these sections in EPRM that either relate directly to this knowledge area or provide background information.

- Section 17.14: Capacitors
- Section 17.19: Real Power and the Power Factor
- Section 17.20: Reactive Power
- Section 17.22: Complex Power and the Power Triangle
- Section 33.15: Synchronous Motor/Reactive Generator

NCEES Handbook: To prepare for this knowledge area, familiarize yourself with these sections and this figure in the *Handbook*.

- Power Factor Correction and Voltage Support

• Complex Power Triangle (Inductive Load)

The following equations and figures are relevant for knowledge area 4.A.3, Power Factor Correction and Voltage Support.

Change in Reactive Power to Change the Power Angle

Handbook: Power Factor Correction and Voltage Support

$$\Delta P_{\text{reactive}} = P_{\text{real}}(\tan \Phi_1 - \tan \Phi_2)$$

The *Handbook* equation is a combination of equations which can be found in EPRM Chap. 17. The variable P_{reactive} is rare and confusing. The standard reactive power symbol is Q, as seen in the following equation.

$$\Delta Q = P(\tan \phi_1 - \tan \phi_2)$$
$$= P\left(\tan \frac{Q_1}{P_1} - \tan \frac{Q_2}{P_2}\right)$$

The angle ϕ (or Φ) is the power angle, which is equal to the impedance angle and the power factor angle.

The *Handbook* equation calculates the change in the reactive power of a system based on the changes in the power factor angle. *Handbook* figure Complex Power Triangle (Inductive Load) shows the relationship of reactive to real power within the power triangle. EPRM Fig. 17.11 also shows this relationship for both leading and lagging conditions.

Figure 17.11 *Power Triangle*

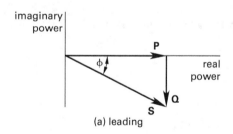

(a) leading

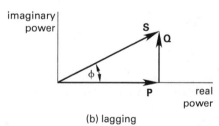

(b) lagging

Since most large loads, such as motors, are coils of wire and inductive (lagging) in nature, most systems are inductive with positive reactive power.

To compensate and bring the power factor closer to one, negative (leading) reactive power is normally added to the system. In doing so, the reactive power is passed between two loads instead of from the load to the generating source and load.

Capacitance Required to Change the Reactive Power

Handbook: Power Factor Correction and Voltage Support

$$C = \frac{\Delta P_{\text{reactive}}}{2\pi f V_{\text{line}}^2}$$

The *Handbook* equation follows from a combination of equations from EPRM Sec. 17.5, Sec. 17.14, Sec. 17.15, and Sec. 17.22.

$$Q = \frac{V^2}{X_c} = \frac{V^2}{\frac{-1}{\omega C}}$$
$$= -V^2 \omega C$$
$$C = -\frac{Q}{V^2 \omega} = -\frac{Q}{V^2 2\pi f}$$

Incorporating the negative sign into the change of reactive power gives the following equation.

$$C = \frac{\Delta Q}{2\pi f V^2}$$

The variable P_{reactive} is rarely used and confusing. The standard reactive power variable is Q. The designer must also keep the sign correct. Recall the negative sign for capacitance reactance is hidden in the X_c value itself.

EPRM Eq. 17.53, Eq. 17.56, Eq. 17.61, and Eq. 17.62 may be helpful in subsequent derivations.

$$P = IV \cos \phi_{\text{pf}} \qquad \textit{17.53}$$
$$= IV \, \text{pf}$$

$$Q = IV \sin \phi_{\text{pf}} \qquad \textit{17.56}$$

$$\text{pf} = \frac{P}{S} \qquad \textit{17.61}$$

$$\frac{X}{R} = \tan(\arccos \text{pf}) \qquad \textit{17.62}$$

Power Factor Loss Reduction

Handbook: Power Factor Correction and Voltage Support

$$\% \text{ loss reduction} = \left(1 - \left(\frac{\text{pf}_1}{\text{pf}_2}\right)^2\right) \times 100\%$$

The *Handbook* equation simply provides a metric to measure the improvement of the power factor for a given change in reactive load.

4. POWER QUALITY

Power quality is the measure of the degree to which the voltage, frequency, and the given waveforms of a power supply system conform to the established requirements and standards. For transmission systems supplying most households and businesses, this standard is ANSI C84.1.

Knowledge Area Overview

Key concepts: These key concepts are important for answering exam questions in knowledge area 4.A.4, Power Quality.

- power quality problems and their effects on different devices
- effects of harmonics on power systems and calculating total harmonic distortion
- percent voltage unbalance of a three-phase supply
- troubleshooting techniques using preventative tips

PE Power Reference Manual (EPRM): Study these sections in EPRM that either relate directly to this knowledge area or provide background information.

- Chapter 20: Transient Analysis
- Chapter 23: Generation Systems
- Appendix 20.A: Resonant Circuit Formulas

NCEES Handbook: To prepare for this knowledge area, familiarize yourself with these sections and figures in the *Handbook*.

- Power Quality

The following equations are relevant for knowledge area 4.A.4, Power Quality.

Voltage Unbalance

Handbook: Power Quality

$$\text{voltage unbalance}(\%) = \frac{\text{maximum voltage deviation}}{\text{average voltage}} \times 100\%$$

The *Handbook* equation is a definition provided by the National Electrical Manufacturers Association (NEMA), Standard MG-1. It is used as a metric for the measure of the voltage imbalance in a three-phase system.

Imbalance can result in increased losses, malfunctioning equipment, and increased wear on generators, as well as other impacts.

Total Harmonic Distortion (THD)

Handbook: Power Quality

$$\text{THD}(\%) = \frac{\sqrt{\sum_{n=2}^{\infty} V_n^2}}{V_1} \times 100\%$$

The numerator represents the rms values of the n_{th} harmonics summed. The denominator represents the fundamental harmonics. For a 60 Hz system, this value is the voltage at 60 Hz.

THD is limited by various standard or equipment requirements and is generally around 3%.

EPRM Eq. 23.11 takes into account all of the harmonics and includes a generic variable that can be used for any DC signal base.

$$X_{\text{rms}} = \sqrt{X_{\text{dc}}^2 + \sum_{n=1}^{\infty} X_n^2} \qquad \textit{23.11}$$

EPRM Sec. 23.11 includes in-depth information on harmonics, including a table of harmonic frequencies.

Resonant Frequency of an *RLC* Circuit

Handbook: Power Quality

$$\omega_r = \frac{1}{\sqrt{LC}}$$

EPRM: Sec. 20.11

$$\omega_0 = 2\pi f_0 = \frac{1}{\sqrt{LC}} \qquad \textit{20.30}$$

The resonant frequency is normally shown with the zero subscript. These equations are only applicable to an *RLC* circuit.

Series Resonant Circuit Quality Factor

Handbook: Power Quality

$$Q = \frac{\omega_r L}{R} = \frac{1}{\omega_r R C}$$

EPRM: Sec. 20.11

$$
\begin{aligned}
Q &= \frac{X}{R} = \frac{\omega_0 L}{R} = \frac{1}{\omega_0 R C} \\
&= \frac{1}{R}\sqrt{\frac{L}{C}} = \frac{\omega_0}{(\text{BW})_{\text{rad/s}}} \\
&= \frac{f_0}{(\text{BW})_{\text{Hz}}} \\
&= G\,\omega_0 L \\
&= \frac{G}{\omega_0 C}
\end{aligned}
\qquad 20.32
$$

The *Handbook* and EPRM Eq. 20.32 are equivalent. The subscripts r and 0 are both used to indicate "resonant". These equations are only applicable to an *RLC* circuit.

EPRM Eq. 20.25 can also be used to determine the quality factor.

Parallel Resonant Circuit Quality Factor

Handbook: Power Quality

$$Q = \frac{R}{\omega_r L} = \omega_r R C$$

EPRM: Sec. 20.12

$$
\begin{aligned}
Q &= \frac{R}{X} = \omega_0 R C = \frac{R}{\omega_0 L} = R\sqrt{\frac{C}{L}} \\
&= \frac{\omega_0}{(\text{BW})_{\text{rad/s}}} = \frac{f_0}{(\text{BW})_{\text{Hz}}} = \frac{\omega_0 C}{G} = \frac{1}{G\omega_0 L}
\end{aligned}
\qquad 20.37
$$

The *Handbook* and EPRM equations are equivalent. These equations are only applicable to an *RLC* circuit.

EPRM Eq. 20.25 can also be used to determine the quality factor.

Series or Parallel Resonant *RLC* Circuit Bandwidth

Handbook: Power Quality

$$F_B = \frac{\omega_r}{Q}$$

The *Handbook* equation uses the symbol F_B for the bandwidth. F is used since this is the frequency response of the circuit. The subscript B stands for bandwidth.

EPRM Eq. 20.25 is used to determine the quality factor, Q, and is a rearrangement of the *Handbook* equation, though the two equations use different variables. In the EPRM equation, the variable for resonant frequency is f_0 rather than ω_0, and the variable for bandwidth is BW instead of F_B.

$$
\begin{aligned}
Q &= 2\pi\left(\frac{\text{maximum energy stored per cycle}}{\text{energy dissipated per cycle}}\right) \\
&= \frac{f_0}{(\text{BW})_{\text{Hz}}} = \frac{\omega_0}{(\text{BW})_{\text{rad/s}}} \\
&= \frac{f_0}{f_2 - f_1} = \frac{\omega_0}{\omega_2 - \omega_1} \begin{bmatrix}\text{parallel} \\ \text{or series}\end{bmatrix}
\end{aligned}
\qquad 20.25
$$

5. FAULT CURRENT ANALYSIS

Fault current analysis is highly technical, and several complicated computer programs are used to analyze systems. A simple method can be used to obtain first-order results. This is called the MVA method and is explained in detail in the EPRM Chap. 26. An important concept to realize when performing such analysis is that once a fault occurs, the energy of momentum in a motor is transferred to the fault; that is, a motor becomes a generator—a source to the fault—during fault conditions.

Knowledge Area Overview

Key concepts: These key concepts are important for answering exam questions in knowledge area 4.A.5, Fault Current Analysis.

- types of electrical faults
- sources of fault current
- generator behavior during a fault
- fault current contribution calculations
- functions of overcurrent protective devices

PE Power Reference Manual **(EPRM):** Study these sections in EPRM that either relate directly to this knowledge area or provide background information.

- Chapter 26: Power Distribution

NCEES Handbook: To prepare for this knowledge area, familiarize yourself with this section in the *Handbook.*

- Fault Current Analysis

The following equation is relevant for knowledge area 4. A.5, Fault Current Analysis.

Short-Circuit MVA for a Balanced 3-Phase Fault

Handbook: Fault Current Analysis

EPRM: Sec. 26.9

$$S_{sc} = \sqrt{3}\, V_{nom} I_{sc}$$

The worst-case fault current is found using the MVA method. The fault current, I_{sc}, is the current that flows in a system or component for a unit voltage from an infinite source. The fault apparent power, S_{sc}, and fault current are used to calculate the maximum fault power and current.

6. TRANSFORMER CONNECTIONS

The primary and secondary windings of three-phase transformers can be connected as either delta or wye. Wye connections, with a common neutral, offer economic and operating advantages. Delta connections allow third-harmonic voltages common to all transformers to circulate within the delta so that line electrical parameters are unaffected. Transformers normally have at least one delta-connected winding.

Knowledge Area Overview

Key concepts: These key concepts are important for answering exam questions in knowledge area 4.A.6, Transformer Connections.

- transformer parameters from open-circuit test results

- transformer parameters from short-circuit test results

- operation of zigzag transformers

***PE Power Reference Manual* (EPRM):** Study these sections in EPRM that either relate directly to this knowledge area or provide background information.

- Chapter 27: Power Transformers

NCEES Handbook: To prepare for this knowledge area, familiarize yourself with these sections and figures in the *Handbook*.

- 3-Phase Circuits

The following equations, tables, and figures are relevant for knowledge area 4.A.6, Transformer Connections.

Three-Phase Transformer Connections

Handbook: 3-Phase Circuits

EPRM: Sec. 27.5

$$V_L = V_P \quad [\text{delta}]$$
$$I_L = \sqrt{3}\, I_P \quad [\text{delta}]$$

$$V_L = \sqrt{3}\, V_P \quad [\text{wye}]$$
$$I_L = I_P \quad [\text{wye}]$$

For delta connections, the line and phase voltage and current are given by EPRM Eq. 27.8 and Eq. 27.9, respectively.

For wye connections, the line and phase voltage and current are given by Eq. 27.10 and Eq. 27.11, respectively.

When voltage and current ratios are given, they are assumed to specify the line conditions, regardless of the connection. When only a turns ratio is given, the line quantities must be found. EPRM Fig. 27.3 displays line-phase relationships.

See EPRM Fig. 27.10 for a diagram of delta and wye connections of three single-phase transformers connected as a three-phase transformer.

Figure 27.3 *Transformer Connections*

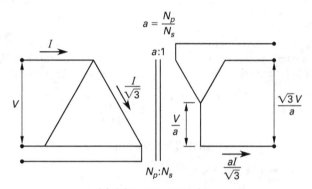

(a) delta-wye connection

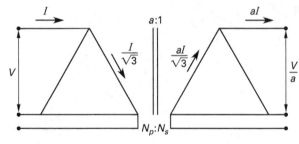

(b) delta-delta connection

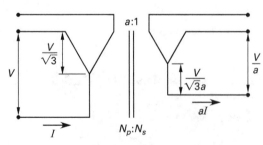

(c) wye-wye connection

Transformer Schematic

EPRM: Sec. 27.5

A standard transformer schematic is shown in EPRM Fig. 27.2.

Figure 27.2 *Standard Transformer Schematic*

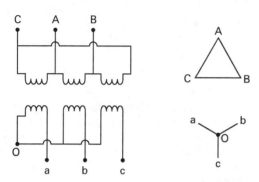

Though no equations are shown in the *Handbook*, recognizing transformer connections determines the electrical equations used in analysis.

Buck Transformers/Boost Transformers/ Autotransformers

EPRM: Sec. 27.14

An autotransformer has only one winding common to both the primary and secondary circuits. EPRM Fig. 27.15 shows typical autotransformer connections.

Figure 27.15 *Autotransformer Connections*

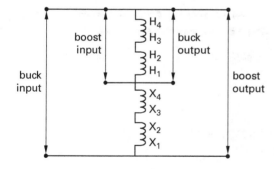

These are sometimes referred to as a buck or boost transformer, depending on the connection. Connections between the primary and secondary windings are made on one winding to either boost the input, which raises the output, or buck the input, which lowers the output.

7. TRANSMISSION LINE MODELS

Transmission lines are generally separated into three types depending upon the length of the line: short, medium, and long.

Knowledge Area Overview

Key concepts: These key concepts are important for answering exam questions in knowledge area 4.A.7, Transmission Line Models.

- conductor types and their characteristics
- the unit of the circular mil
- the diameter of a wire from its nominal size
- properties and types of stranded conductors
- stranding of different types of stranded conductors

PE Power Reference Manual (**EPRM**): Study these sections in EPRM that either relate directly to this knowledge area or provide background information.

- Section 15.4: Characteristic Impedance
- Section 16.3: Resistance
- Section 19.9: Resistance
- Section 27.4: Voltage Regulation
- Chapter 28: Transmission Lines

NCEES Handbook: To prepare for this knowledge area, familiarize yourself with these sections and figures in the *Handbook*.

- Transmission Line Models
- Short Transmission Line Model
- Medium Transmission Line Model
- Long Transmission Line Model
- Transmission Line Parameters
- Line Inductance and Inductive Reactance

The following equations and figures are relevant for knowledge area 4.A.7, Transmission Line Models.

Line Impedance in Ω Per Phase Per Unit Length

Handbook: Transmission Line Models

EPRM: Sec. 28.12–Sec. 28.14

$$z = r + jx$$

z is the line impedance in ohms per phase per unit length. Normally when a quantity is per unit length, it is given the subscript l or L. However, another usage is to make the variables lowercase, or use a combination of lowercase variables and subscripts. Be cautious of the units.

Equivalent Total Impedance of the Line

Handbook: Transmission Line Models

EPRM: Sec. 28.12

$$Z = (z)(\text{length}) = R + jX_L$$

The terms represent total amounts but are calculated per unit length to correct for conductor spacing and then multiplied by the transmission line length.

Equivalent Total Admittance of the Line

Handbook: Transmission Line Models

$$Y_c = (y)(\text{length})$$

EPRM: Sec. 28.14

$$
\begin{aligned}
Y_l &= jB_{C,l} \\
&= \frac{j}{X_{C,l}} \\
&= -\frac{1}{jX_{C,l}} \\
&= j\omega C_l
\end{aligned}
\tag{28.43}
$$

In the EPRM equation, the admittance, Y, per unit length is equal to susceptance, with j accounted for in the expression jB. Both have units of siemens, S, which is the reciprocal of ohms, Ω^{-1}.

Line Propagation Constant

Handbook: Transmission Line Models

EPRM: Sec. 28.17

$$
\begin{aligned}
\gamma &= \sqrt{zy} \\
\gamma &= \alpha + j\beta
\end{aligned}
$$

The propagation constant, γ, is in units of radians per meter (or mile) and measures the speed of the electromagnetic wave within the material.

α is called the attenuation constant measured in nepers per meter (or mile). A neper is a logarithmic unit for ratios. It makes it mathematically easier to use decibels when determining the signal loss in a line. β is the phase constant measured in radians per meter (or mile).

The name "constant" is a bit of a misnomer given the equation represents the change of a signal (wave) from one port to the next, both in magnitude and phase.

Line Characteristic Impedance

Handbook: Transmission Line Models

EPRM: Sec. 28.1

$$Z_c = \sqrt{\frac{z}{y}}$$

The *Handbook* equation and EPRM Eq. 28.1 are equivalent. If the transmission line is terminated with the characteristic impedance, no power is reflected back to the source. If the termination impedance is not Z_c, then the power (signal) from the generator (source) will be reflected back to the generator.

The characteristic impedance of free space can be found using EPRM Eq. 15.4.

$$Z_0 = \frac{|\mathbf{E}|}{|\mathbf{H}|} = \frac{E}{H} = \sqrt{\frac{\mu_0}{\epsilon_0}} = \mu_0 c \tag{15.4}$$

In a medium with known relative permeability and relative permittivity, the surge impedance can be found using EPRM Eq. 15.3.

$$Z = \sqrt{\frac{\mu}{\epsilon}} = \sqrt{\frac{\mu_r \mu_0}{\epsilon_r \epsilon_0}} = Z_0 \sqrt{\frac{\mu_r}{\epsilon_r}} \tag{15.3}$$

Phase Voltage and Current Between the Sending and Receiving Ends

Handbook: Transmission Line Models

EPRM: Sec. 28.11

$$
\begin{aligned}
V_s &= A V_r + B I_r \\
I_s &= C V_r + D I_r
\end{aligned}
$$

Transmission lines and other two-port devices, such as transistors, can be represented using ABCD parameters, shown in matrix form in EPRM Fig. 28.3.

Figure 28.3 *Transmission Line Two-Port Network*

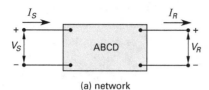

(a) network

$$V_S = AV_R + BI_R$$
$$I_S = CV_R + DI_R$$

(b) equations

$$\begin{bmatrix} V_S \\ I_S \end{bmatrix} = \begin{bmatrix} A & B \\ C & D \end{bmatrix} \begin{bmatrix} V_R \\ I_R \end{bmatrix}$$

(c) matrix form of equations

EPRM Table 28.3 shows the ABCD parameters for transmission lines of various lengths.

Transmission Efficiency of the Line

Handbook: Transmission Line Models

EPRM: Sec. 28.11

$$\eta = \frac{P_r}{P_s} \times 100\%$$

The design of all transmission lines must account for voltage regulation and efficiency. The subscript r indicates the receiving end. The subscript s indicates the sending end. The *Handbook* equation and EPRM Eq. 28.32 are equivalent.

Line Percentage Voltage Regulation

Handbook: Transmission Line Models

EPRM: Sec. 28.11; Sec. 27.4

$$\%\text{VR} = \frac{|V_{r,\text{nl}}| - |V_{r,\text{fl}}|}{|V_{r,\text{fl}}|} \times 100\%$$

$$\%\text{VR} = \frac{\left|\dfrac{V_s}{A}\right| - |V_{r,\text{fl}}|}{|V_{r,\text{fl}}|} \times 100\%$$

The first *Handbook* equation and EPRM Eq. 28.31 are equivalent. The second *Handbook* equation and EPRM Eq. 27.7 are equivalent. While they may differ only in the subscripts used, both equations express the voltage regulation for transformers in terms of the rated values and the turns ratio.

Approximate Line Power Transfer

Handbook: Transmission Line Models

$$P \cong \frac{V_s V_r}{X_L} \sin \delta$$

The *Handbook* equation is a single-phase power transfer formula. It is very similar to a standard formula with the exception that the δ is the angle between the sending- and receiving-end voltages.

Line Maximum Power Transfer

Handbook: Transmission Line Models

$$P \cong \frac{V_s V_r}{X_L}$$

The maximum power transfer occurs when no angular difference exists between the sending and receiving ends.

When only reactance is considered, EPRM Eq. 19.6 follows the *Handbook* equation.

$$P = IV = I^2 R = \frac{V^2}{R} \qquad \text{19.6}$$

Surge Impedance Loading

Handbook: Transmission Line Models

$$\text{SIL} = \frac{V_{\text{line}}^2}{Z_s}$$

The *Handbook* equation calculates the surge impedance loading, SIL, in units of megawatts (MW). This is only the case if the voltage is in units of kilovolts (kV).

The surge impedance, Z_s, is equivalent to the characteristic impedance, which is shown in EPRM Eq. 15.3.

$$Z = \sqrt{\frac{\mu}{\epsilon}} = \sqrt{\frac{\mu_r \mu_0}{\epsilon_r \epsilon_0}} = Z_0 \sqrt{\frac{\mu_r}{\epsilon_r}} \qquad \text{15.3}$$

Line Constants for the Short Transmission Line Model

Handbook: Short Transmission Line Model

EPRM: Sec. 28.11; Sec. 28.12

$$A = D = 1$$
$$B = Z$$
$$C = 0$$

Short transmission lines have a line length ≤ 50 miles. EPRM Eq. 28.34 and Eq. 28.35 give the voltage and current at any point along the line.

$$V_S = V_R + I_R Z \qquad \text{28.34}$$

$$I_S = I_R \qquad \text{28.35}$$

EPRM Table 28.3 shows the ABCD parameters for transmission lines of various lengths.

Line Constants for the Medium Transmission Line Model

Handbook: Medium Transmission Line Model

EPRM: Sec. 28.11; Sec. 28.13

$$A = 1 + \frac{ZY}{2}$$

$$B = Z$$

$$C = \left(1 + \frac{ZY}{4}\right)Y$$

$$D = A$$

Medium transmission lines have a line length > 50 miles and ≤ 150 miles. The *Handbook* provides the line constants only for the nominal π-model, not the nominal T-model. EPRM Eq. 28.38 and Eq. 28.39 give the voltage and current at any point along the line.

$$V_S = \left(1 + \tfrac{1}{2}YZ\right)V_R + \left(Z\left(1 + \tfrac{1}{4}YZ\right)\right)I_R \qquad \text{28.38}$$

$$I_S = YV_R + \left(1 + \tfrac{1}{2}YZ\right)I_R \qquad \text{28.39}$$

EPRM Table 28.3 shows the ABCD parameters for transmission lines of various lengths.

Line Constants for the Long Transmission Line Model

Handbook: Long Transmission Line Model

EPRM: Sec. 28.11; Sec. 28.14

$$A = \cosh\left(\gamma l\right)$$

$$B = Z_c \sinh\left(\gamma l\right)$$

$$C = \frac{\sinh\left(\gamma l\right)}{Z_c}$$

$$D = A$$

Long transmission lines have a line length > 150 miles. EPRM Eq. 28.49 and Eq. 28.50 give the voltage and current at any point along the line.

$$
\begin{aligned}
V &= \tfrac{1}{2}V_R(e^{\gamma x} + e^{-\gamma x}) \\
&\quad + \tfrac{1}{2}I_R Z_0(e^{\gamma x} - e^{-\gamma x}) \\
&= V_R \cosh\gamma x + I_R Z_0 \sinh\gamma x
\end{aligned}
\qquad \text{28.49}
$$

$$
\begin{aligned}
I &= \tfrac{1}{2}I_R(e^{\gamma x} + e^{-\gamma x}) \\
&\quad + \tfrac{1}{2}\left(\frac{V_R}{Z_0}\right)(e^{\gamma x} - e^{-\gamma x}) \\
&= I_R \cosh\gamma x + \left(\frac{V_R}{Z_0}\right)\sinh\gamma x
\end{aligned}
\qquad \text{28.50}
$$

EPRM Table 28.3 shows the ABCD parameters for transmission lines of various lengths.

Conductor DC Resistance

Handbook: Transmission Line Parameters

EPRM: Sec. 16.3

$$R_{\text{DC}} = \frac{\rho l}{A}$$

The resistivity, ρ, length, l, and cross-sectional area, A, of a substance determines its resistance.

Conductor AC Resistance

Handbook: Transmission Line Parameters

EPRM: Sec. 28.4

$$R_{\text{AC}} = kR_{\text{DC}}$$

The skin effect ratio for AC resistance, k, determines the effective resistance. The *Handbook* and EPRM Eq. 28.11 are equivalent. EPRM Table 28.2 shows various skin effect ratios.

Table 28.2 Skin Effect Ratios

x	K_R[a]	$K_{L,\text{int}}$[b]
0.0	1.00000	1.00000
0.5	1.00032	0.99984
1.0	1.00519	0.99741
2.0	1.07816	0.96113
4.0	1.67787	0.68632
8.0	3.09445	0.35107
16.0	5.91509	0.17649
32.0	11.56785	0.08835
50.0	17.93032	0.05656
100.0	35.60666	0.02828
∞	∞	0

[a]K_R is the skin effect ratio for AC resistance.
[b]$K_{L,\text{int}}$ is the skin effect ratio for internal inductance.

Line Resistance at a Different Temperature

Handbook: Transmission Line Parameters

$$\frac{R_2}{R_1} = \frac{M + T_2}{M + T_1}$$

The *Handbook* equation allows you to calculate losses and energy changes as temperature varies over a day or over the seasons. *M* is the line's temperature constant.

EPRM Eq. 19.3 can be used in a manner similar to the *Handbook* equation. The resistivity, ρ, is in units of $\Omega \cdot m$.

$$\rho = \rho_{20}\big(1 + \alpha_{20}(T - 20°\text{C})\big) \qquad 19.3$$

Average Line Inductance

Handbook: Line Inductance and Inductive Reactance

EPRM: Sec. 28.9

$$L = \frac{\mu_0}{2\pi}\ln\frac{D_{\text{eq}}}{\text{GMR}}$$

$$D_{\text{eq}} = \sqrt[3]{D_{\text{ab}}D_{\text{bc}}D_{\text{ca}}}$$

The first *Handbook* equation and EPRM Eq. 28.24 are used to determine the average line inductance, per unit length. The second *Handbook* equation and EPRM Eq. 28.23 are equivalent. The variables D_e and D_{eq} are equivalent and represent the equilateral spacing of a fully transposed three-phase line.

The GMR is the geometric mean radius, which depends upon the configuration of the cables. See the section on Geometric Mean Radius for more information.

Geometric Mean Radius

Handbook: Line Inductance and Inductive Reactance

$$\text{GMR [one bundle] } r' = 0.7788r$$

$$\text{GMR [two bundle]} = \sqrt{r'd}$$

$$\text{GMR [three bundle]} = \sqrt[3]{r'dd}$$

$$\text{GMR [four bundle]} = \sqrt[4]{r'dd\sqrt{2d}}$$

EPRM: Sec. 28.7

$$\text{GMR} = re^{-1/4} \qquad 28.20$$

The GMR is the geometric mean radius, which depends upon the configuration of the cables. The geometric mean radius is also called the self geometric mean

difference. While r is the radius of a single conductor, the r can be replaced with a r', indicating the GMR (the self geometric mean difference) for each of the conductors in a one-, two-, three- [equilateral], or four-conductor [square] bundle. The variable d is the actual distance between the conductors. From the *Handbook*, the single cable value for the GMR is $0.7788r$, which correlates with the value for the GMR in EPRM Eq. 28.20.

See EPRM Fig. 28.2 for various transmission line spacings.

Average Inductive Reactance

Handbook: Line Inductance and Inductive Reactance

EPRM: Sec. 28.10

$$X_L = (2.022 \times 10^{-3})f\ln\frac{D_{\text{eq}}}{\text{GMR}}$$

The *Handbook* equation is given in units of ohm-miles (Ω-mi). The *Handbook* equation and EPRM Eq. 28.27 are equivalent.

Average Capacitance to Neutral

Handbook: Line Inductance and Inductive Reactance

$$C = \frac{2\pi\epsilon_0}{\ln\dfrac{D_{\text{eq}}}{\text{GMR}_c}}$$

EPRM: Sec. 28.9

$$C_l = \frac{2\pi\epsilon_0}{\ln\dfrac{D_e}{r}} \qquad 28.25$$

$$= \frac{5.56 \times 10^{-11}}{\ln\dfrac{D_e}{r}} \quad [\text{in F/m}]$$

EPRM Eq. 28.25 is applicable to a three-phase circuit. EPRM Eq. 28.22 is applicable to a single-phase circuit.

$$C_l = \frac{\pi\epsilon_0}{\ln\dfrac{D}{r}} = \frac{2.78 \times 10^{-11}}{\ln\dfrac{D}{r}} \quad [\text{in F/m}] \qquad 28.22$$

Average Capacitive Reactance

Handbook: Line Inductance and Inductive Reactance

EPRM: Sec. 28.10

$$X_c = \frac{1.779}{f} \times 10^6 \ln\frac{D_{eq}}{\text{GMR}_c}$$

While there are minor differences in the *Handbook* equation and EPRM Eq. 28.29, they are equivalent. EPRM Eq. 28.29 uses a value of 1.781 in the numerator of the first term on the right side of the equation. The equations also differ in units, hence the conversion factor of 10^6.

8. POWER FLOW

Power flow analysis, or load flow analysis, is fundamental to power system operation and planning.

Knowledge Area Overview

Key concepts: These key concepts are important for answering exam questions in knowledge area 4.A.8, Power Flow.

- power flow in a transmission line
- power flow studies
- power flow reference circuit analysis
- real and reactive power flow

PE Power Reference Manual **(EPRM):** Study these sections in EPRM that either relate directly to this knowledge area or provide background information.

- Section 19.31: Two-Port Networks
- Section 26.7: Fault Analysis: Symmetrical
- Section 28.11: Transmission Line Representation
- Section 28.17: High-Frequency Transmission Lines
- Section 29.2: Power Flow

NCEES Handbook: To prepare for this knowledge area, familiarize yourself with these sections and figures in the *Handbook*.

- Power Flow

The following equations and figures are relevant for knowledge area 4.A.8, Power Flow.

Power Flow Between Two Balanced Three-Phase Voltage Sources

Handbook: Power Flow

$$P = 3\left(\frac{E_1 E_2}{X}\right)\sin\delta$$

The *Handbook* equation indicates the power flow from the sending-end source voltage, E_1, to the receiving-end voltage, E_2, with an intervening reactance, X.

δ is the angle of E_1 with respect to E_2. Therefore, E_2 is considered the reference.

Even though E_1 and E_2 are used to describe two voltage sources, power is flowing from source 1 to source 2 with the reactive impedance X between them.

Bus Admittance Matrix

Handbook: Power Flow

$$\mathbf{I} = \mathbf{Y}_{\text{bus}}\mathbf{V}$$

In complex computer software designed to model electrical distribution systems, matrices for each bus are created using the estimated (design) values of the wiring and bus bars. Using these matrices for each bus, short circuit analysis can occur for a multitude of faults. EPRM Eq. 19.62 and Eq. 19.63 show the matrix in Eq. 19.61 broken out into individual equations.

$$\begin{bmatrix} i_1 \\ i_2 \end{bmatrix} = \begin{bmatrix} y_{11} & y_{12} \\ y_{21} & y_{22} \end{bmatrix}\begin{bmatrix} v_1 \\ v_2 \end{bmatrix} \qquad \text{19.61}$$

$$i_1 = y_{11}v_1 + y_{12}v_2 \qquad \text{19.62}$$

$$i_2 = y_{21}v_1 + y_{22}v_2 \qquad \text{19.63}$$

Although EPRM Fig. 26.7 illustrates a two-port network, it is an example of the admittance matrix results. The types of faults are also shown.

Figure 26.7 *Fault Types, from Most Likely to Least Likely*

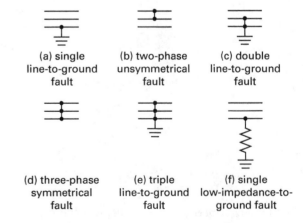

(a) single line-to-ground fault

(b) two-phase unsymmetrical fault

(c) double line-to-ground fault

(d) three-phase symmetrical fault

(e) triple line-to-ground fault

(f) single low-impedance-to-ground fault

Elements of Bus Admittance Matrix

Handbook: Power Flow

$$Y_{ij} = \begin{cases} y_i + \sum_{k=1, k\neq i}^{N} y_{ik} & [\text{for } i = j] \\ -y_{ij} & [\text{for } i \neq j] \end{cases}$$

There are a limited number of bus types in a given network used in fault analysis software.

The term Y_{ij} represents the total admittance of the bus. The term y_i is the admittance from bus i to the reference bus, of which there is only one. The other terms are the admittances to buses j and k.

The variable N is the total number of buses in the network.

9. POWER SYSTEM STABILITY

Power system stability involves the study of the dynamics of a power system after having been subject to a disturbance. The dynamics of a power system are associated with the interactions among its components.

Knowledge Area Overview

Key concepts: These key concepts are important for answering exam questions in knowledge area 4.A.9, Power System Stability.

- inertia constant calculation
- power angle equation
- requirements for power system stability
- types of stability studies that can be performed

PE Power Reference Manual **(EPRM):** Study these sections in EPRM that either relate directly to this knowledge area or provide background information.

- Section 17.22: Complex Power and the Power Triangle
- Section 29.11: IEEE Brown Book
- Section 30.16: Stability
- Section 32.10: Generator and Motor Action
- Section 32.12: Torque and Power
- Section 33.12: Synchronous Motors
- Section 41.14: Measurement Standards and Conventions

NCEES Handbook: To prepare for this knowledge area, familiarize yourself with this section in the *Handbook.*

- Power System Stability

The following equations and figure are relevant for knowledge area 4.A.9, Power System Stability.

Swing Equation

Handbook: Power System Stability

$$J \frac{d^2\theta_m}{dt^2} = T_a = T_m - T_e$$

During steady-state conditions, the acceleration (second derivative) of the angular displacement of the rotor mass, θ_m, in units of radians (mechanical) is equal to zero. Note the mass m in the *Handbook* equation should be a subscript.

J is the moment of inertia, a property of all masses, in units kg·m^2.

The result of multiplying the first two terms is the net accelerating torque, T_a.

The net accelerating torque results from subtracting the net output, which is the electromagnetic torque developed by the generator, T_e, after losses, from the input torque, which is developed by the mechanical input to the rotor shaft, T_m. All torques are in units of N·m.

The relationship between torque and power is shown in EPRM Eq. 32.16.

$$T_{\text{N·m}} = \frac{9549 P_{\text{kW}}}{n_{\text{rev/min}}} = \frac{1000 P_{\text{kW}}}{\omega_{\text{mech}}} \qquad 32.16$$

In a generator, a torque is applied to the shaft causing a relative motion between the conductors in the stator and the rotor, which carries the magnetic field. This is called generator action. Once there is relative motion, a voltage is induced in the stator windings resulting in current flow and the generation of a separate magnetic field that interacts with the first. The current flow is the desired output.

In a generator, generator action (the desired result) results in motor action (the undesired result). The motor action is a result of the conservation of energy. A constant input force or torque is required to continue generating an electrical output. For more in-depth information, review EPRM Sec. 32.10.

Formulation of the Swing Equation Utilizing Per-Unit Quantities

Handbook: Power System Stability

$$\frac{2H}{\omega_s}\frac{d^2\delta}{dt^2} = P_a = P_m - P_e$$

The *Handbook* equation is an alternate form of the "swing equation" expressed in terms of per-unit power.

δ is the rotor angle in units of electrical radians.

H is a machine constant defined in terms of the stored energy in the machine at any given time, which will impact its ability to stabilize.

Machine *H* Constant

Handbook: Power System Stability

$$H = \frac{\text{stored kinetic energy at synchronous speed (MJ)}}{\text{machine rating (MVA)}}$$

$$= \frac{\frac{1}{2}J\omega_{sm}^2}{S_{mach}}$$

The variable for angular frequency, ω_{sm}, in the equation represents the mechanical speed in units of mechanical radians per second.

The greater the energy stored, the more difficult to destabilize and the longer to return to a stable condition once disturbed.

This is the same stored motor energy that, during a fault, results in the motor becoming a generator and delivering its energy to the fault point.

Power Angle Equation

Handbook: Power System Stability

$$P(\delta) = \frac{|E||V_t|}{X}\sin\delta$$

The power angle equation relates the generated voltage to the terminal voltage and the reactance. Recall that reactance dominates resistance due to the many coils of wire in a generator or motor.

The δ is the angle between the two voltages. EPRM Fig. 33.9 shows the relationship between the rotor fields and resultant fields.

Figure 33.9 *Magnetic Field Generated by a Synchronous Generator*

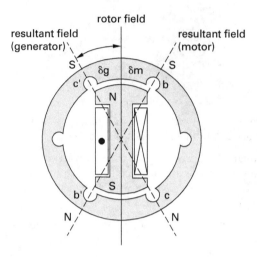

Critical Clearing Angle

Handbook: Power System Stability

$$\delta_{cr} = \cos^{-1}\big((\pi - 2\delta_0)\sin\delta_0 - \cos\delta_0\big)$$

$$t_{cr} = \sqrt{\frac{4H(\delta_{cr} - \delta_0)}{\omega_s P_m}}$$

The critical clearing angle is the maximum change in load angle before clearing a fault without a loss of synchronism.

The corresponding time is the "time to clear," which becomes the design parameter for the protection system.

If a synchronous machine loses synchronism, the currents generated can be extremely large and can often result in the catastrophic failure of the machine.

10. DEFINITIONS

autotransformer: A transformer with only one winding that is common to both the primary and secondary circuits associated with that winding.

bus: Akin to a node in electrical circuits, a bus in a power system is the vertical line at which the components of the power system are connected. Buses in a power system are associated with four quantities: magnitude of the voltage, the phase angle of the voltage, active or true power, and the reactive power.

characteristic impedance: The equivalent resistance of a transmission line if it were infinitely long.

critical clearing angle: The maximum change in load angle before clearing a fault without a loss of synchronism.

fault: An abnormal electric current. A short-circuit fault happens when the current bypasses the normal load. An open-circuit fault occurs when a circuit is interrupted by a failure.

geometric mean radius: The geometric mean distance between strands of a conductor, or between conductors. It presupposes an external flux linkage.

ideal independent voltage source: A two-terminal circuit element where the voltage across the terminals is maintained regardless of the current flowing through the terminals.

machine H constant: The ratio of kinetic energy stored at the synchronous speed to the generator kVA or MVA rating; also called the inertia constant.

neper: A logarithmic unit for the ratios of units.

power angle: The phase angle difference between the applied voltage and the generated emf. Also known as torque angle.

power factor: The ratio of real power to apparent power measured in kilovolt-amperes.

power quality: A measure of the degree to which voltage, frequency, and the waveform of a power supply system correspond to the required specifications.

propagation constant: A measure of the change in a sinusoidal electromagnetic wave in terms of amplitude and phase, while propagating through a medium.

reactive power: The product of the rms values of the current and voltage multiplied by the quadrature of the current.

resonance: A condition of frequency in a circuit where the inductive reactance and capacitive reactance cancel one another—where they are equal but of opposite sign.

RLC circuit: An electrical circuit consisting of a resistor (R), an inductor (L), and a capacitor (C), connected in series or in parallel.

skin effect: The tendency of an AC current to flow near the surface of a conductor and whose current decreases exponentially with increasing depth into the conductor.

surge impedance loading: The power load in which the total reactive power of the line becomes zero.

total harmonic distortion: A measurement of the harmonic distortion present in a signal. It is defined as the ratio of the sum of the powers of all harmonic components to the power of the fundamental frequency.

voltage regulation: The measure of the change in voltage magnitude between the sending and receiving end of a component or system.

voltage support: The ability to absorb or support the reactive power of a system, thus maintaining the proper voltage level—that is, maintaining voltage within code limits.

X/R ratio: Ratio of reactance to resistance critical to the response of a system during fault conditions. The greater the ratio, the longer it takes to return to a steady-state condition.

11. NOMENCLATURE

A	ABCD parameter	–
A	area	m^2
B	ABCD parameter	–
B	susceptance	S
BW	bandwidth	Hz
c	speed of light	km/s or miles/hr
C	ABCD parameter	–
C	capacitance	F
d	diameter	m
d	distance	m
D	ABCD parameter	–
D	diameter, distance	m
E	induced voltage	V
f	frequency	Hz
F	frequency response	Hz
G	conductance	S
GMR	geometric mean radius	m
H	high side transformer connections	–
H	machine constant	–
I	DC or rms current	A
J	moment of inertia	kg·m^2
k	proportionality constant	–
k, K	skin effect ratio	–
L	inductance	H
M	temperature constant for a given material	–
n, N	number	–
n	speed	rpm
P	power	W
pf	power factor	–
Q	quality factor	
Q	reactive power	VAR
r	radius	m
r	real (resistive)	Ω
R	resistance	Ω
S	apparent power	VA
SIL	surge impedance loading	VA
T	temperature	°C

T	torque	N·m
THD	total harmonic distortion	%
V	voltage	V
VD	voltage drop	V
x	instantaneous reactance	Ω
X	reactance	Ω
y	admittance per unit length	S/m
Y	admittance	S
z	impedance	Ω
Z	impedance	Ω

P	phase	
r	receiving	
R	receiving resistance, or resonant	
rms	root mean square	
s	secondary, sending, synchronous	
S	sending	
sc	short circuit	
sm	synchronous machine	
t	terminal	

Symbols

α	attenuation constant	Np/m
α	temperature coefficient	1/°C
β	phase constant	radians/m
γ	propagation constant	radians/m
δ	torque angle	radians
ϵ	permittivity	F/m
η	efficiency	%
θ	angle, angular displacement	radians (mechanical)
μ	permeability	H/m
π	pi	
ρ	resistivity	Ω·m
ω	angular frequency	radians/s
Φ	power angle	radians

Subscripts

0	free space, original (unloaded), or resonant
a	accelerating
B	bandwidth
c	characteristic, conductor
C	capacitive
cr	critical
dc	direct current
e	electromagnetic, equivalent
eq	equivalent
fl	full load
i	two-port current
l	line
L	inductive, line
LN	line to neutral
m	mechanical
n	number
nl	no-load
nom	nominal
p	primary

9 Protection

Exam specification 4.B, Protection makes up between 14% and 21% of the PE Electrical Power exam (between 11 and 17 questions out of 80).

The organization of this chapter follows the order of knowledge areas given by the NCEES for this exam specification. Each knowledge area is covered in the following numbered and sections.

Content in blue refers to the *NCEES Handbook*.

Content in red is additional essential information.

The topic of protection is not specifically listed in the *Handbook* with the exception of lightning protection and the Codes and Standards exam specification coverage. Even so, one would expect future updates to the *Handbook* to expand this coverage given the requirements coverage.

Even without explicit coverage in the *Handbook*, a professional engineer should have a breadth of knowledge in the topic of protection. To that end, this chapter will focus on this topic and where such information exists in the *PE Power Reference Manual* (EPRM). Focus will also be on the widely used *National Electrical Code* (NEC) for households and buildings and the *National Electrical Safety Code* (NESC) for transmission and distribution systems.

Please refer to Chap. 3, Codes and Standards, in this book for in-depth information about the NEC and the NESC.

1. OVERCURRENT PROTECTION

Overcurrent is any current over the rating of the equipment or the ampacity of the conductors. Overcurrent results from overload, short circuit, or ground fault. Overcurrent protection devices used to protect the equipment from faults include circuit breakers, reclosers, sectionalizers, and fuses.

To solve problems on protection, review the key ideas and equations covered in Chap. 7, Electric Power Devices, in this book. To understand the key ideas in further detail, review the *PE Power Reference Manual* (EPRM) Chap. 18.

Knowledge Area Overview

Key concepts: These key concepts are important for answering exam questions in knowledge area 4.B.1, Overcurrent Protection.

- behavior of protection system components
- familiarity with the NEC and its applicability
- grounding of a power system
- overcurrent conditions and protection system components
- overcurrent protection process
- resources on overcurrent protection in the NEC
- role of branch circuits in a protection system

PE Power Reference Manual (EPRM): Study these sections in EPRM that either relate directly to this knowledge area or provide background information.

- Section 26.4: Overcurrent Protection
- Section 31.2: Power System Grounding
- Section 31.3: Power System Configurations
- Section 31.8: Protection System Elements
- Section 44.4: Wiring and Protection
- Section 44.5: Wiring and Protection: Grounded Conductors
- Section 44.6: Wiring and Protection: Branch Circuits
- Section 44.7: Wiring and Protection: Feeders
- Section 44.8: Wiring and Protection: Branch Circuit, Feeder, and Service Calculations
- Section 44.9: Wiring and Protection: Overcurrent Protection
- Section 44.10: Wiring and Protection: Grounding

Codes and standards: These are the most important sections of the codes and standards for this knowledge area.

- NEC Art. 280: Surge Arresters, Over 1000 Volts

- NEC Art. 285: Surge-Protective Devices (SPDs), 1000 Volts or Less

NCEES Handbook: The *Handbook* does not include any sections for this knowledge area.

The following figure, tables, and key ideas apply to knowledge area 4.B.1, Overcurrent Protection.

Protection System Components

EPRM: Sec. 31.8

A protection system consists of three main elements: transducers, relays, and breakers. When a fault occurs, a transducer sends a signal to a relay which operates a breaker. This is discussed in more detail in the Protective Devices knowledge area in this chapter.

Power System Configurations

EPRM: Sec. 31.3

Radial bus configuration. A radial system is a serially connected system with a single source. It is economical to build but less reliable because a problem at any point in the network affects all downstream loads. A radial system is also easier to protect because a fault current can only flow from the source to the fault, making it less complex to calculate. The fault current is independent of the generating capacity as the system is remote from the source.

Network bus configuration. A network system is more complicated to protect than a radial system. A fault current can occur from multiple sources and from multiple directions. Because the source may be close to the fault, the fault current may vary widely depending on location.

Substation configuration. Substation configurations are discussed in detail in EPRM Sec. 31.3 and Chap. 1, Measurement and Instrumentation, of this book.

Power System Grounding

EPRM: Sec. 31.2

The grounding of a power system influences the fault current levels. Protective relays must sense the level of ground-fault current before a ground fault can damage the power system. In a true ungrounded system, there is zero fault current. In a real system, capacitive coupling of the feeders to ground results in a ground path if a fault occurs. The ungrounded system with a ground fault is shown in EPRM Fig. 31.2.

Figure 31.2 Ungrounded System with Ground Fault

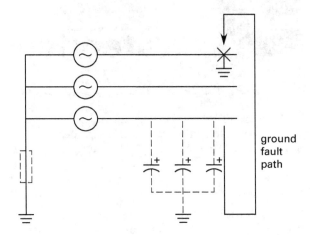

A disadvantage of ungrounded systems is that when a ground fault occurs on one phase, the other two phase voltages become $\sqrt{3}$ larger than their normal values. The phasor representation of a grounded phase is shown in EPRM Fig. 31.3.

Figure 31.3 Phasor Diagram of Grounded Phase

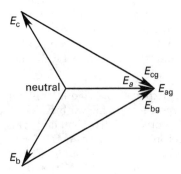

On low-voltage systems, the insulation levels of conductors are usually set based on lightning-induced phenomena, which generally produce larger voltage increases than those that are fault induced. On high-voltage systems, fault-induced voltages are more critical, so a solidly grounded neutral is used.

Branch Circuits

EPRM: Sec. 44.6

A branch circuit consists of circuit conductors between the final overcurrent device protecting the circuit and the outlet. Branch circuits are covered in NEC Art. 430. See Chap. 3, Codes and Standards, in this book for more in-depth information. The cross-reference for articles relating to branch circuits for other special conditions is given in NEC Table 210.2.

NEC Wiring Protection and Grounding

EPRM: Sec. 44.9; Sec. 44.10

In addition to coverage in EPRM, overcurrent protection and grounding are discussed in NEC Art. 240 and Art. 250.

NEC Art. 240 describes the overcurrent protection requirements and the electrical devices that accomplish them. NEC Table 240.3 lists specific types of equipment and the applicable article containing the overcurrent requirements. NEC Table 240.3 is a useful starting point for determining overcurrent requirements.

Conductors are protected in accordance with their ampacities [NEC Sec. 240.4]. Standard ampere ratings for fuses and circuit breakers are given in NEC Sec. 240.6 as 15, 20, 25, 30, 35, 40, 45, 50, and so on. Thermal devices not designed to interrupt short circuits cannot be used for overcurrent protection for short circuits or grounds [NEC Sec. 240.9]. The remainder of NEC Art. 240 discusses the location of overcurrent devices, tap requirements, fuses, and circuit breakers.

NEC Art. 250 describes grounding and bonding requirements for electrical systems.

Electrical systems are connected to the earth in a way that will limit voltage from lightning strikes, line surges, or other transients [NEC Sec. 250.4(A)(2)]. Electrical systems are also grounded in order to maintain the voltage at a stable level relative to ground to ensure connected equipment is subjected to a maximum potential difference set by the designer. Equipment is grounded as well, the primary goal being the safety of people in contact with the equipment. The overall requirements of the fault current path and their organization in the NEC are given in NEC Sec. 250.4(B)(4), while specific equipment requirements are cross-referenced in NEC Table. 250.3, which is a good starting point for finding grounding information.

A ground-fault circuit interrupter (GFCI) from EPRM Fig. 44.4 is shown.

Figure 44.4 Ground-Fault Circuit Interrupter (GFCI)

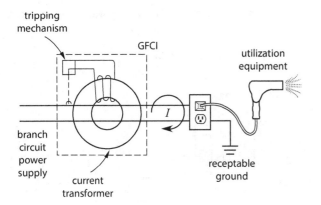

An imbalance in one of the lines passing through the current transformer due to part of the current passing through a ground downstream of the GFCI trips the mechanism much faster than an upstream overcurrent protection device would. This is the operating principle that allows for protection of personnel.

See Chap. 3, Codes and Standards, in this book for additional coverage on the NEC as it pertains to the exam.

2. PROTECTIVE RELAYING

The purpose of relaying is to remove any element that is functioning abnormally from a given power system. It prevents further damage, minimizes danger to personnel, and stabilizes the system.

Knowledge Area Overview

Key concepts: These key concepts are important for answering exam questions in knowledge area 4.B.2, Protective Relaying.

- characteristics of overcurrent relays
- fundamentals of directional, ratio, differential, overcurrent, and pilot protection
- major types of protective relays
- operating time of a relay
- protection zones of a power system
- relay current settings for faults in a transformer with differential protection
- relay reliability

PE Power Reference Manual **(EPRM):** Study these sections in EPRM that either relate directly to this knowledge area or provide background information.

- Chapter 31: Protection and Safety

NCEES Handbook: To prepare for this knowledge area, familiarize yourself with this section in the *Handbook*.

- Single-Line Diagrams

The following figures, table, and key ideas and figures are relevant for knowledge area 4.B.2, Protective Relaying.

ANSI/IEEE Devices

Handbook: Single-Line Diagrams

EPRM: Table 31.1

A list of some of the device numbers for common power system elements is given in the *Handbook*, per ANSI/IEEE C37.2. Refer to EPRM Table 31.1 for a more comprehensive list of device numbers and their functions.

Relay Types

EPRM: Sec. 31.10

A summary of the relays described in detail in EPRM Sec. 31.10 is presented in alphabetical order.

Differential relay. A differential relay operates when currents at the input and output of a zone are not identical. This type of relay is effective in isolating faults.

Directional relay. This type of relay responds to faults either to the left or to the right of its location in the direction of normal operation. It does not trip if fault current flows in the other direction.

EPRM Fig. 31.16(b) shows the equation characteristics of a directional relay.

Overcurrent relay. This type of relay is also called a level detection relay. An overcurrent relay operates to trip a circuit breaker when the fault current exceeds a given value. The properties of a typical level detection relay are shown in EPRM Fig. 31.14.

Figure 31.14 *Level Detector Characteristics*

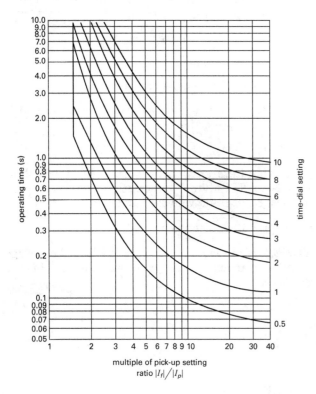

The current required to actuate the relay is called the pick-up current. The pick-up setting is the setting at which the relay operates, which is the ratio of the fault current to the pick-up current.

The time dial is the delay introduced in the relay operating time. The time dial setting and the pick-up setting are selected to give the desired time response in the event of fault.

Phase angle relay. This type of relay compares the current and voltage to determine the direction of power flow. A phase angle relay detecting a 90° impedance angle can react under the assumption that a fault has occurred.

Pilot relay. This type of relay sends signals from transmission lines to relays at the terminals of the lines. Pilot relays provide fast protection and are used on long transmission lines.

Ratio relay. This type of relay is also called a distance relay or an impedance relay. A ratio relay responds to faults based on the impedance within a certain distance of the relay location. The ratio relay depends on the ratio of the voltage to the current. The relay can be adjusted for any desired distance, assuming standard impedance per unit distance. It senses the impedance and trips if the impedance becomes lower than the reference. EPRM Fig. 31.16(a) shows the equation characteristics of a ratio relay.

Relay Reliability

EPRM: Sec. 31.5

A relay's reliability is the probability that it will perform its intended function under a given set of environmental or electrical conditions. The reliability requirements of protection system relays include dependability and security. The relay is

- dependable if there is a measure of certainty that it will operate for designed faults

- secure if there is a measure of certainty that it will not operate incorrectly

Zones of Protection

EPRM: Sec. 31.6

The protection zones of a power system are shown in EPRM Fig. 31.11.

Figure 31.11 *Closed and Open Zones*

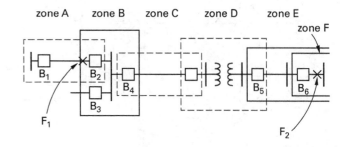

A power system is divided into zones of protection for reliability and minimization of interruption of service. Zones can be closed or open. A closed zone, also called a differential, unit, or absolutely selective zone, is one where all power apparatuses entering the zone are monitored at the entry points. In an open zone, also called an unrestricted, non-unit, or relatively selective zone, monitoring does not occur, and the zone is defined by the level of fault current.

3. PROTECTIVE DEVICES

A protective device is a device used to control, protect, or isolate electrical equipment from fault.

Knowledge Area Overview

Key concepts: These key concepts are important for answering exam questions in knowledge area 4.B.3, Protective Devices.

- appropriate circuit breakers for an electrical system

- function of a crowbar circuit

- functions of different types of protective devices, such as fuses, circuit breakers, sectionalizers, and reclosers

- role of the current transformer as a protective device

- selection of an appropriate protective fuse for an electrical system

PE Power Reference Manual **(EPRM):** Study these sections in EPRM that either relate directly to this knowledge area or provide background information.

- Section 26.4: Overcurrent Protection

- Section 26.12: IEEE Red Book

- Chapter 31: Protection and Safety

- Section 38.5: Protective Devices

- Section 43.7: Shock Protection

- Section 45.6: NESC Part 1

Codes and standards: These are the most important sections of the codes and standards for this knowledge area.

- NESC Part 1, Sec. 19, Surge Arresters: Rules 190–193

NCEES Handbook: The *Handbook* does not include any sections for this knowledge area.

The following figures and key ideas are relevant for knowledge area 4.B.3, Protective Devices.

Protective Devices for Transmission Lines

EPRM: Sec. 31.13

Protective devices in order of complexity and cost are shown in EPRM Fig. 31.19.

Figure 31.19 *Protective Devices*

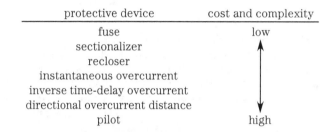

protective device	cost and complexity
fuse	low
sectionalizer	
recloser	
instantaneous overcurrent	
inverse time-delay overcurrent	
directional overcurrent distance	
pilot	high

A *fuse* is a common protective device that comes in a series of standard sizes, each with a time-current characteristic curve. When the current is too high, the fuse element melts and interrupts the current flow.

A *sectionalizer* is a self-contained device used in conjunction with other devices, such as reclosers and circuit breakers, to permanently isolate a faulted portion of a distribution system. It has no interrupting capacity. The sectionalizer counts the number of times a fault is seen and opens the circuit at a predetermined number of counts. The circuit is de-energized by the recloser or circuit breaker.

A *recloser* senses a fault and interrupts power to a system. After a preset time, it attempts to reenergize the system by closing. If other protective devices have isolated the fault, the recloser remains closed.

An *overcurrent device* operates when a fault exceeds a given value.

Pilot protection schemes use a communication channel between the ends of a transmission line that enables the system to clear faults over 100% of the line using overlapping zones and differential relaying.

Non-pilot protection refers to a system that responds only to the parameters at one end of a transmission line.

Fuses, sectionalizers, and reclosers are also discussed in EPRM Sec. 26.4.

Protection System Components

EPRM: Sec. 31.8

The three main elements of a protection system are transducers, relays, and breakers.

Transducers sense the desired signal and convert it into a form usable by the relays. Examples of transducers are current and voltage transformers.

Relays monitor the input from the transducers and provide an output to operate breakers in the event of a fault.

Breakers interrupt power and isolate sections of the power system during a fault.

An uninterruptable power supply (UPS) ensures that electricity continuously reaches crucial equipment. A UPS is used as a source of backup power when the normal power system fails.

Circuit Breakers

EPRM: Sec. 26.4

A circuit breaker is an electromagnetic device used to open and close a circuit. When an overload occurs, the heat causes the breaker to operate by opening a set of electrical contacts. The trip characteristics of some circuit breakers are adjustable. An instantaneous trip is one that occurs without delay. An inverse time trip is one in which a delay is deliberately instituted.

Types of circuit breakers:

- low-voltage circuit breakers

  - less than 1000 V

  - miniature circuit breaker (MCB) up to 100 A

  - molded-case circuit breaker (MCCB) up to 1000 A

- medium-voltage circuit breakers

  - voltages between 1000 V and 69 kV

  - vacuum circuit breakers up to 3000 A

  - air circuit breakers up to 10,000 A

  - SF6 circuit breakers up to 10,000 A

- high-voltage circuit breakers

  - over 69 kV

  - oil, SF6, or vacuum

Crowbar Circuits

EPRM: Sec. 31.14

A crowbar circuit is used with a circuit attached to a power supply to prevent damage due to overvoltage from the power supply. A crowbar example circuit is explained in EPRM Sec. 31.14 and shown in EPRM Fig. 31.20.

Figure 31.20 *Crowbar Example Circuit*

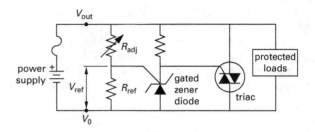

Current Transformer Burden

EPRM: Sec. 31.9

Current transformers (CTs) can be used as instrument transformers or as transducers for protection circuits.

Figure 31.12 *Current Transformer*

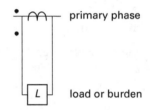

The burden is the amount of power drawn from the circuit connecting the secondary terminals of instrument transformers, usually given as apparent power in volt-amperes. The term burden distinguishes power for operation of the device itself from power sensed, which is being monitored and is referred to as the load. As long as the burden is within the rating of the CT, only acceptably small errors are introduced in the phase relationship between the primary and secondary currents, and the current transformation ratio is exact.

See Chap. 1, Measurement and Instrumentation, in this book for additional information on current transformers.

Surge Arresters (NESC – Part 1, Section 19)

EPRM: Sec. 45.6

Surge arresters are common in the utility industry and are used to protect equipment from switching surges and lightning strikes. Rules for surge arresters are mentioned in NESC Sec. 19, Rules 190–193. The basic requirements, however, come from the IEEE Standard C62.1 and C62.11.

Surge arresters are discussed in detail in EPRM Sec. 38.5. See Chap. 3, Codes and Standards, in this book for additional coverage on the NESC as it pertains to the exam.

4. COORDINATION

The primary principle to remember regarding coordination is that the protective device closest to the fault should respond first and the one farthest from the fault should respond last. This maximizes protection of the equipment while maintaining the availability of the greatest portion of the equipment during an emergency.

Knowledge Area Overview

Key concepts: These key concepts are important for answering exam questions in knowledge area 4.B.4, Coordination.

- development of recommended protective device settings using a coordination study

- generator capability/control/protection coordination

- inverse time relationship of the protection relays

- overcurrent relay coordination

PE Power Reference Manual **(EPRM):** Study these sections in EPRM that either relate directly to this knowledge area or provide background information

- Section 31.6: Zones of Protection

- Section 31.7: Relay Speed

- Section 31.17: Generator Capability/Control/Protection Coordination

- Section 31.18: IEEE Buff Book

- Section 31.21: IEEE Blue Book

NCEES Handbook: The *Handbook* does not include any sections for this knowledge area.

The following figures and key ideas are relevant for knowledge area 4.B.4, Coordination.

Coordination Study

EPRM: Sec. 31.18; Sec. 31.21

A short-circuit study and a coordination study are performed to achieve the highest possible coordination.

These studies include:

- a one-line diagram showing protective device ampere ratings and associated designations, cable sizes and lengths, transformer kVA and voltage ratings, motor and generator kVA ratings, and switchgear/switchboard/panelboard designations

- descriptions, purpose, basis, and scope of the study

- tabulations of the worst-case calculated short-circuit duties as a percentage of the applied device rating (automatic transfer switches, circuit breakers, fuses, etc.); the short-circuit duties should be upward-adjusted for X/R ratios that are greater than the device design ratings

- protective device time versus current coordination curves with associated one-line diagram identifying the plotted devices, tabulations of ANSI protective relay functions, and adjustable circuit breaker trip unit settings

- fault study input data, case descriptions, and current calculations, including a definition of terms and guide for interpretation of the computer printout

- comments and recommendations for system improvements, where needed

The protective device closest to the fault should operate first. If not, the next device should operate, and so on. Use the time-current characteristic (TCC) curve of each device in the power system to achieve coordination among protective devices. Separation of the TCC lines indicates coordination. Where lines cross, there is a loss of coordination.

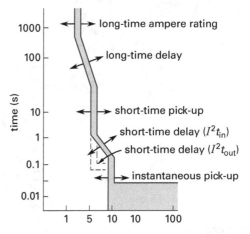

Relay Speed

EPRM: Sec. 31.7

Protection relays have an inverse time relationship. That is, there is an inverse relationship between the time it takes for the relay to make a decision and the security of that decision.

An instantaneous relay operates as soon as the logic circuitry makes a decision. A time-delay relay has an intentional time delay inserted after a decision is made. A high-speed relay operates in a specified time, such as 50 ms, or three cycles of a 60 Hz system, as prescribed in various electrical standards.

Overcurrent Relay Coordination

EPRM Sec. 31.6; Sec. 31.18; Sec. 31.21

The principle of timed overcurrent protection is that relays are applied appropriately at the terminal of zones, and each relay is then given both a pick-up and a time delay setting. The current pickup establishes the sensitivity of the relay. The time-delay overcurrent relay is used for coordination with other protective relays or devices.

The primary relays that are closest to the fault should trip first, while the more remote backup relays wait to see if the fault is cleared. If not, then the backup relays time out and send a trip signal to the breaker.

Factors influencing the time delay:

- the operating time of the circuit breakers; allow 0.1 s for this

- overtravel to account for moving part inertia in the circuit breaker of about 0.1 s

- a margin for uncertainties: 0.1 to 0.3 s

The total coordination time margin is about 0.3 s.

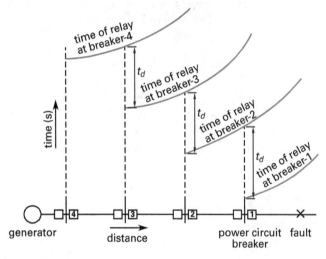

In the figure, t_d represents the time delay (in seconds) between relay operations.

Generator Capability/Control/Protection Coordination

EPRM: Sec. 31.17

The capability of the generating unit, the limits of the excitation control system, and the generator protection must be coordinated to avoid equipment damage and power outages.

5. DEFINITIONS

breaker: An automatically operated electrical switch.

burden: The load on an instrument transformer when at rated value. This load is normally given in units of VA and must be maintained within allowable limits to ensure the accuracy of the instrument transformer.

continuous duty: Operation at a constant load for an indefinitely long time.

fuse: An electrical safety device that operates to provide overcurrent protection primarily through melting the metal wire or strip inside the device.

overcurrent: Any current in excess of rated values for the equipment or conductors. Overcurrent can be due to overload, short circuit, or ground fault.

overload: Operation of equipment in excess of normal full load rating, or of a conductor in excess of its ampacity. An overload is not a short-circuit or ground fault.

periodic duty: Operation in which load conditions are regularly recurrent.

recloser: Interrupts fault current and has the capability of reclosing the circuit multiple times automatically.

relay: An electrically operated switch operated by signals from outside sources.

sectionalizer: Isolates the faulted section of a system. It does not interrupt the fault current.

transducer: A device for transforming energy from one form to another for use in an electrical circuit.

varying duty: Operation at loads, and for intervals of time, both of which may be subject to wide variation.

6. NOMENCLATURE

B	breaker	–
E	voltage	V
F	fault	–
I	current	A
L	load	–
R	resistance	Ω
t	time	s
V	constant or rms voltage	V

Subscripts

0	ground point
a	line, phase, or phasor
adj	adjustable
b	line, phase, or phasor
c	line, phase, or phasor
d	delay
f	fault
g	ground
L	load
p	pickup
r	relay
ref	reference

Index

INDEX - P